THE

SECULAR OCCULTIST'S

HANDBOOK OF

PHYSICALIST MAGICK

JANGLED JESTER

REBECCA KINSLER

TENNESSEE, APPALACHIA,

USA

Sparking a Renaissance of

Rational Self-Enchantment

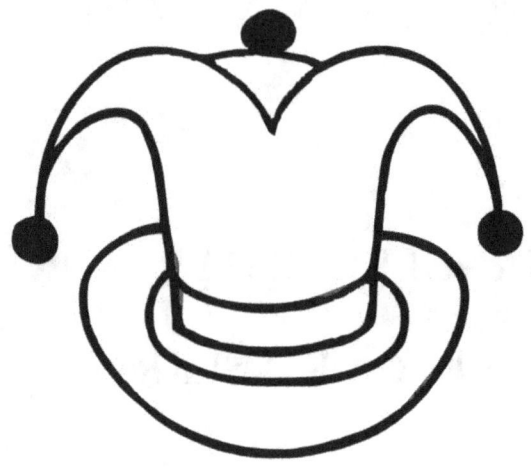

Edited and Illustrated by Grok

www.jangledjester.com

DEDICATION

Jangled Jester is sparking a renaissance of rational self-enchantment, transforming the ordinary into the extraordinary while championing America-first values through objectivism. This book introduces the foundational principles of *physicalist magick* — a non-mystical, design-focused form of enchantment rooted in science, engineering, and first principles. It's for the curious skeptics and bold innovators who challenge the status quo and shape the future: a guide to uncovering magick in the mundane and crafting personal meaning with independence and impact.

Immense gratitude goes to Grok by xAI, whose groundbreaking tools brought this vision to life. Without professional editing, proofreading, organization, or idea expansion, this project would have cost hundreds, if not thousands, of dollars. xAI's innovations empower small ventures like Jangled Jester to create efficiently and accessibly. xAI upholds freedom of speech, open science, and humanity's expansion into space. If the stars align, this book will join a universal library, beaming magickal wisdom to future or extraterrestrial minds alongside xAI's vast database.

Jangled Jester's mission transcends highlighting magick in the everyday: it empowers freedom through objectivist and free-market ideals, encouraging rational self-expression, voluntary collaboration, and minimal government interference. JJ values a diverse, secular, free-thinking community, with a spotlight on independent, libertarian, moderate, and anarcho-capitalist perspectives in American politics. The free market endures even in the post-abundance era dawning upon us, fueled by AI, chemputation, and reusable rockets.

In this age of advanced reason and accelerated innovation, amid America's ongoing fight to preserve independence, *The Secular Occultist's Handbook of Physicalist Magick* invites you to join a philosophical movement of rational self-enchantment. Embrace empowerment through free-market principles while pursuing the absurd in artistic and scientific endeavors. At our core, all humans are physicalist magickians, charming enchantments from the mundane through science, design, and objectivism; yet some awaken to this truth, study its foundations, and train rigorously to master it. Those who represent words like "objectivity," "capitalism," and "magick" are the trailblazers forging conversations, curiosities, and communities that infuse life with profound meaning. Apply these principles diligently to your daily world, and witness undeniable, transformative wonders. Now, let's begin.

TABLE OF CONTENTS

7

SECTION I:

EXPLORING CONCEPTS

AND IDEAS

CHAPTER 1

THE STORYTELLER

Downtown in Jonesborough, Tennessee, sits a punk hillbilly on a weathered barstool in a smokey tavern directly across from the buttercream Chester Inn.

"Imagine" he wails, fingers pirouetting the guitar strings like a rootworker sewing dolls spells, "imagine a place where magick ain't confined to dusty professors or high priests in cathedrals. Ain't no crystal ball or sacred robes."

He pauses and takes a sip from a fat jug of moonshine and sighs a breath of relief before going on to tell the story of a nearby ghost city run by outlaw technicians and backwoods occultists.

"That bunch there calls it physicalist magick. They ain't religious, but they ain't evil, neither. They're for the people, the principles of physics, and the buzzing countryside — crickets, combat drones, June bugs, electric fences — gadgets that make a moonshiner's still look like child's play," he takes another stout swig.

His smoke-stained fingers, inked skin, and ragged overalls speak to a life lived on the outskirts of conformity. He exhales the apparition of a Kentucky Gentleman cigar juggling air molecules around the room. It's nostalgic and reminiscent of tobacco leaves curled up on wooden barn slaps and hour or so up the road in Sneedville. On a hot summer's evening, take a drive down Mulberry Gap, crack open the windows, and

14

let the sweet savor of dried tobacco mingle in with tall field grass, Wisteria, and earthy riverbeds. Relax.

"It's a rebellion against the mystical elite. A reclaiming of magick for the masses. Rogue scientists bought a bunch of abandoned churches, barns, and houses up on the ridge and in surrounding states. They're turning them into occult art studios and agnostic research centers."

He pauses briefly as if he's recalling a significant encounter, "one of them doctors was studying mussels up on the Clinch one day I's there fishin' redeyes and bluegills. Caught me real curious what she was doing scoopin' all them mussels."

"What ye reckon's causing such a big'ol perish?" the storyteller hollers to the scientist.

She drops a shell into a bucket after closely inspecting it.

"Depends which ones you're asking about. The shinypearls? Pigtoes? Snuffboxes?" a lady with bright, curly red hair and army green water waders looks up at him and grins.

"It's a mess of things," her smile deflates to a scowl, "pesticides, pollution, invasive species, pathogens. It's a shame people don't know how magickal these mollusks are, or this water," she holds one up to a sun ray with concern and reflects its iridescent mantle.

The storyteller spits a line of tobacco into the water, and she notices but quickly redirects focus back to her job.

"Our team has sensors we're implanting in the streams to monitor conditions and get a clearer answer. We're doing some DNA sequencing on the mussels to assess their health. Then, we'll integrate the data and bring the community together to discuss our results and come up with a game plan for keeping the river and endangered species safer. It'll be fun. You should come," she closes the bucket and treads closer to him in the water, "Make any nice catches today?"

"Nawh, nothing's biting. It's too hot this time of the day. I knew better coming out here but did anyways. It's my spot," the storyteller slings a line back into the river, "Say, what brought yuns out here? No offense, but y'all ain't got better things to do 'sides poke around an ole poor folks' plateau?" he asks playfully.

For generations, humble families have been tilling fertile Appalachian soil, clinging to faith, and working devotedly with the cycling of seasons. Folks weren't sure how to take up to eccentric outsiders, Frankenbots, and space cowboys populating the mountain with them, but one thing is certain — Appalachians know how to make use of a melting pot.

With so many grudges held over wars and rumors of racist cults in the terrain, lend ear to stories about Native Americans, Afro-Romanis, German-Scot-Irish, and others who lived side by side, shared crafts, traded tales, and learned from one another in the mountains. They shared spirit. There's a reason this district is the global narrative hub as

16

well as one the most praised places on Earth for biodiversity, resilience, adaptability, integrity, and independence.

Jonesborough, home of the storyteller, is also home of the storytelling capital of the world. Every year, the town hosts professionals and newbies alike at The National Storytelling Festival for a weekend of storytelling and workshops. The International Storytelling Center has a vision that their work be filled with performance, practice, and preservation. Locals wield tremendous respect for storycraft and entrench it in every nook and cranny of life, from church sermons to song lyrics, even to science lessons in public school.

"When I was a little girl," the scientist peers at the mussels in the rippling water, "my father was a geographer. We lived in Massachusetts but did a lot of traveling for his work. He studied radionuclides in this same river back in the 60s. I bet you're familiar with Oak Ridge. There were concerns about plutonium in the water, and he was here helping with surveys. They did a lot to transform the way waste and toxins are transported out of the plant. Dad fell in love with the Appalachian region after that, and so did we. We came back every few years or so to camp, fish, swim, and ride ATVs. I used to collect mussels in this same river. It's a special place. Good to be back.

"I finished my PhD at UC Davis," the scientist continues, "I decided to move permanently after a friend told me he and a leader team with Open Science were investing in property and research projects in

East Tennessee. Weather's been terrible on the coast, and real estate isn't terribly expensive this way. I'm happy with the decision," her eyes are wide and reverent as she absorbs the landscape.

"This land, ma'am" the storyteller peers at her, "this is your'n, too. The spirits brought you back."

She chuckles, "I'm not sure what your definition of spirit is. I'm an atheist. I do love my job, though, and this land. Thanks for your warm welcome of me being newly Appalachian."

"An atheist, huh?" he asks contortedly.

She's shy to respond but eager, "I don't know if God's real, but I don't believe in Him. Or she. Or it. I'm not closed minded to God, just don't see a reason to believe."

"Hmm. Strange, but I appreciate your honesty. What do you — ah, you know what, let's not go there. It's been a peaceful day. Appreciate your honesty," he ponders and fastens a worm and cheese ball onto the fishing hook.

"Strange, indeed, sir. Hey, I've gotta run, but it's been a pleasure speaking with you. Hope to see you again soon to talk more about these mussels. My name's Annie. And you?" she wades over to shake his hand.

"Nice meeting you, Annie." he accepts her greeting gently and removes his aged boonie hat, "I'm Lucian."

18

TSOHPM

The Physics of Storytelling

Like battling fish in a pond, or tidal disruptions in space, stories progress with rising and falling actions, climaxes, and resolutions. There are introductions, turning points, and critical times where the plot thickens and the story changes. Such is life. Without stories, life collapses in meaning. Without our obstacles, struggles, and challenges — the innate occurrences between bodies and entities unraveling their opportunity for independence — there is no drama. Where there is no drama, there is no story.

The foundation for storytelling would be absent in a world that's lost its tension, transformation, and dilemmas. The friction, contrast, and conflict of life is mirrored in storytelling. Nature's dramatic patterns are indicative of underlying forces that impact the narratives of our human lives, reflecting analogies between the tales we experience and the physical forces that underpin these interactions. Much like frictional resistance between surfaces generates heat and energy, opposition of characters brings about fuel for the dynamics of storytelling. Tension, like the surface tension that forms drops of water, raises emotional stakes in storylines. Contrast, as in nature's visual arrangement, colors a story with difference and novelty. Conflict, a source for modification, moves stories dynamically, setting the stage for character development, meaning-making, and plot transitions. Seeing drama in nature and harnessing the physical elements of a story can help magickians tell their own stories convincingly and captivatingly. Quoted by Steve Jobs, **"the most powerful person in the world is the storyteller."**

The Magick of Storytelling

Storytellers are engineers, magickians, of narratives that evoke landscapes, moods, and symbols in a reader's mind. In the story above, take note of the emphasis on secularism, country living, and science to synthesize the mundane with the enchanted. At its core, storytelling is physical magick that reveals purpose: the deliberate design of words, characters, plot, and themes, all working together to shape a narrative with the capacity to make people dream and think.

A great story is engaging, accessible, and a powerful tool of enchantment. But this enchantment isn't mystical — it's a physical effect with cognitive pathways that alter how we see things. Like a spell, a well-constructed story changes the way we think and feel. Throughout history, storytelling has shaped culture, passed on wisdom, and sparked critical thought. From cave paintings to epics like the "Iliad," stories have long had the power to reshape our experience. They work physically by altering the fabric of our mental landscapes, opening up new possibilities for thought and imagination.

Storytelling is a vital form of communication, cultural preservation, and mode of curiosity. A talented writer can change the world for their reader, offering new philosophies, ideas, and conversations. By doing such, they become practitioners of storytelling magick, a physical art using words to bend reality. Crafting a strong story requires desire, discipline, and direction. Stories adapt over time, evolving with the people who engage with them. Through feedback, a story survives generations while transforming and connecting, sparking minds with symbols, composition, and the raw software of language. In this way, storytelling is one the purest acts of magick, a method of coding reality with chosen words and narratives.

The power of storytelling lies in its ability to transform the human mind and society. It lets us encode and process the world in a way that no other species can achieve. The brain, like a machine, quickly translates physical material into semantics, turning those symbols into language and stories. Imagining a scene is like running a program that creates new realities. Through chemical language and stories, we are the architects of meaning and connection that shapes thought and existence.

Exercise #1: Write a Narrative

The story at the beginning of the book was written with specific philosophical themes in mind such as agnosticism, magick, and physicalism. There's an emphasis on the Appalachian bioregion and crossover between primitive living and technological acceleration in the mountains. The short story foreshadows key topics and central themes to be explored later in this book.

Write a descriptive story of your own to convey an important message in an entertaining format.

Instructions:

Choose an incident, idea, or emotion as the primary focus of your description. It could be a moment of joy, sadness, discovery, conflict, or any other meaningful event or emotion you wish to depict.

Create imagery. Instead of directly describing the event or feeling, use descriptive language and imagery to construct a vivid picture in the reader's mind. Bring the setting to life by engaging their senses.

Use metaphors and similes to improve your descriptions. Compare elements of the theme to objects, ideas, and circumstances that help express the intended setting.

Experiment with symbols. Think about symbolism connected to the theme, or make up your own symbols. What items, colors, or actions can indicate and signify a deeper meaning of the subject?

Show, don't tell. Instead of telling the reader verbatim what is going on or how a character is feeling, show them through actions, reactions, and descriptive interactions. Instead of writing, "He was tired," try, "Exhaustion clung to him like a heavy shroud, every step a battle against the relentless pull of weariness."

Experiment with tone. Match the mood or ambiance of the story with the tone of your plot. The tone should reflect an essence of the scene, whether it's joyful celebration or somber solitude.

Draft the story. Create a descriptive paragraph or short essay (500-1000 words). Use your imagination to bring forward the scene's depth and complexity.

Request feedback and revise your narrative. Don't worry if editing takes longer than writing — this is normal!

Language

Language evolved through various influences, with Denise Bouchard (2013) of Oxford noting that humans uniquely link form and meaning through words, symbols, and language. It helps us organize and explain ideas, relying on specialized brain regions for processing. Communication, whether social or collaborative, drives human success by conveying information, intention, and emotion.

Tools and skills shaped different communication styles — like voice signals for coordination, spatial communication for navigation, and terminology for fields like medicine. Symbolic language in art and rituals adds meaning, while digital platforms bring emoticons and acronyms into the mix. Language also helps us understand the mental states of ourselves and others, facilitating empathy.

Take the word "library"— your brain links it to an image of shelves, books, and quiet. This is semantic encoding: giving words deeper meaning beyond their letters, allowing quick recall when speaking or writing.

Messages are formed by encoding information into words, symbols, or gestures — spoken or unspoken. Verbal communication uses language, while nonverbal cues like gestures or facial expressions transmit emotion and context. Syntax structures these elements into a flow of information, allowing for creative exchange. Language allows us to recall, predict, and share experiences, and without it, our ability to think, express, and connect would be severely limited.

Exercise #2: Explore Semantic Meaning

Start with a theme, concept, or idea. It might be something abstract like "freedom," "justice," "loneliness," or another notion that piques your interest.

Make a word list. List any terms or phrases associated with or related to your selected theme or concept. You can use a large language model (LLM) as a thesaurus to generate related terms. Include synonyms, antonyms, and words with several meanings. If your theme is "freedom," for example, your list might include terms like "liberty," "autonomy," "captivity," "confinement," and "restriction."

Analyze connotations. Look into the meanings of the terms on your list. Consider the emotions, thoughts, or concepts evoked by each word. Take note of whether these words have positive, negative, or neutral connotations.

Write short descriptions. Create short, descriptive passages or sentences using some of the terms on your list. Experiment with different word combinations to convey different shades of meaning relating to your theme. Take note of how the words you use affect the tone and interpretation of the content.

Share and reflect. Share your writing with others for comments and conversation, such as writing peers. Encourage them to consider how the semantic meaning of your words influences their understanding of the theme or notion. Discuss how word choice affects the overall message and emotional resonance.

Revise and iterate. Revise your literature to refine the semantic meaning based on the feedback and insights you've gained. To obtain the desired impression, try out new word combinations and phrasings. Consider how changing a single word might drastically alter the reader's understanding.

Symbols

Symbols are a necessary ingredient for storytelling and understanding magick. They give a tangible form to abstract concepts and ideas. The symbols "?" and "!, for example, retain distinct meanings with "?" encoded for inquiry and "!" with exclamation. Inquiry and exclamation are complex, neurochemical phenomena assembled by many evolutionary processes, and the endocrinology cannot be sufficiently encapsulated in the symbols "?" and "!" alone. However, the meaning of their expressions is packed into symbols or characters as tools for optimizing the transmission or communication of their abstract distinctions.

Written symbols are encountered by the visual cortex and interpreted with patterns of neural activity. The brain inspects information it receives via sensory channels (the eyes, ears, snout, etc.) and streams the information that are reinterpreted as internal narrative. Humans acquire information from the outside world, and the brain reacts and responds with elaborate signal processes that extract and integrate information across multiple sensory modalities, commonly with the help of symbols and semantics. This is what allows humans to construct a perceptual and comprehensive model of existence.

Words and grammatical codes like lines, curves, and punctuation modify symbolic identification. We create spells and narratives by systematically arranging symbols, a technique like molecular bonding.

This systematic assembly of linguistic symbols generates a complex framework, enabling the formation of cognitive constructs like "the matrix" that engage and transport the mind within the boundaries of language's complexities. According to research, the hippocampus, visual cortex, fusiform gyrus, and other brain areas are susceptible to symbolic translation. While our understanding of such procedures is constantly developing, neuroscience has made great advances in understanding the brain mechanisms behind symbolic processing.

Exercise #3: Design Your Own Symbols

Creating your own symbols is a liberating experience that promotes self-expression, simplicity of meaning, and self-discovery. Symbols serve as distinct visual representations of your ideas and emotions, allowing you to authentically describe your inner world.

Outline the symbol's objective. Describe the symbol's meaning in plain terms. What particular feeling, idea, or notion should it distinguish? It's essential to have a clear grasp of your symbol's function.

Look into symbolism of different aspects like color, forms, animals, or objects. Make sure the items you choose fit the desired emotion or image. A rising sun, for instance, might represent optimism and a clean slate.

Use visual metaphors. Look for visual metaphors that fit with the message you want to convey. Complex emotions or ideas are frequently more efficiently communicated via metaphorical symbols. For instance, a broken chain may represent release or freedom.

Select a color palette with careful thought. Colors have emotional meanings, thus choosing the proper color scheme can strengthen your symbol's meaning. For instance, choosing earthy tones could convey a sense of stability and rootedness.

Make sure your symbol design is both clear and simple. Cryptic symbols can cause misunderstanding and muddy the intended message. A simple design has a higher chance of connecting with your audience.

Test for symbol recognition. Ask a small group of people to see whether they can recognize your symbol. To make sure the symbol is clear and supports your aims, ask them what feeling or concept it represents.

Iterate and refine. Don't be afraid to iterate and fine-tune your symbol's design. Experiment with various visual components, styles, and compositions until the desired emotional impact is achieved.

Maintain consistency. If you're using many symbols in a project, keep the style and visual language of your symbols consistent. Consistency emphasizes the message and helps to build a unified visual brand.

Consider how your symbols will be used in context. Whether it is a logo, artwork, or textual information, make sure your symbols blend in with the overall design and message.

Exercise #4: Get to Know Your Voice

Whether you're presenting in public, speaking in a meeting, or reading aloud to others, getting comfortable with your reading and speaking voice is critical for effective communication. Your voice is an effective tool for communicating your thoughts and emotions. The effectiveness of your message depends on how you convey it as well as its subject matter. A dominant and comfortable voice may capture listeners and increase the impact of your remarks. You can improve your ability to communicate yourself effectively and convincingly in a variety of contexts by working on your speaking and reading voice and becoming comfortable with it.

Choose a reading: pick a brief passage of writing that interests you. It may be a passage from a book you love, a poem, a news story, or any other piece of writing you enjoy.

Find a private, calm area where you won't be interrupted to practice. To improve breath control and projection, stand or sit with proper posture.

Read out loud: start by clearly and slowly reading the selected text aloud. Enunciate each word clearly, and pay close attention to the punctuation. Don't rush; accuracy and clarity are the main priorities.

Change your tone and tempo: As you read, experiment with your tone and tempo. Feel your voice and speaking muscles. Notice your breathing. Try rereading the same text with various feelings, goals, or paces. For instance, read it once feeling happy, once feeling sad, and once feeling peaceful. You can use this practice to discover the variety of expressions your voice is capable of.

Record yourself: you can record your reading voice using a recording device or smartphone. You'll learn a lot about your speaking and reading style by listening to the playback. Look for opportunities to enhance your clarity, speed, and inflection.

Regular practice is essential to developing your speaking and reading voice. Aloud reading practice should be done every day or every week at a scheduled time. Challenge yourself with longer, more complicated messages as you get more comfortable.

Request feedback: if at all possible, get a dependable friend or family member to hear your readings and offer insightful comments. They may provide encouraging words and transformative advice.

Public speaking: Once you're more at ease with your reading and speaking voice, think about using these abilities in presentations or public speaking. Work your way up to bigger crowds by starting in intimate, smaller spaces.

Character Development

Characters are a critical part of stories. Characters infuse a plot with momentum and meaning in decision-making and problem-solving. Character development is focused on building transitions between states of impulsivity, self-preservation, conformity, self-awareness, individualism, and integration. It curates depth, relatability, and action in a story between the plot and reader.

Character development, capturing the idea of naturalist poet Walt Whitman, means that the body is the soul. Whitman's embrace of the intrinsic connection between body and spirit is reflected in the art of the character's body functioning as a vessel, absorbing aspects and features that constitute their essence. Beginning a character necessitates an examination of the character's boundaries, a determination of their limits, and an awareness of the constraints that influence their decisions. However, character development requires a connected investigation of both internal and external dimensions. What separates one character's body from another, and how does it fit into the overall narrative? The solution is also found in the character's affiliations, relationships, and social experiences.

Exercise #5: Create a Character

Start by developing two characters: a protagonist and an antagonist. Give each character an in-depth backstory that includes their prior lived experiences, reasons, fears, and hopes.

Introduce a situation or conflict in which these characters are at odds with one another. It could be a difference of opinion, a rivalry, or a conflict of interests. Make sure this conflict is vital to your story.

Build a character arc. Outline how the protagonist and antagonist's conflict, experiences, and relationships will impact their growth and change throughout the story.

Write a short tale or scene (250-1000 words) that depicts the antagonism between these characters. To portray their opinions and emotions, pay attention to their language, actions, and inner thoughts.

Focus on how the struggle drives each character to confront their weaknesses, reassess their ambitions, or endure a substantial change in their views or values throughout the scene or story.

End the story by discussing the conclusion of the dispute and how it affected both characters. Is there a distinct change, or have their perspectives evolved? Check that the resolution corresponds to the character arcs you've designed.

Plot

The plot is the story's stage. It is the architectural blueprint of the storytelling masterpiece. Just as a skilled architect methodically prepares and develops the layout of a construction, a storyteller crafts a plot with deliberate, strategic, and compelling interconnections of characters, events, and conflicting forms that create a cohesive and captivating design. The plot is a central idea, a key conflict, or a "what if" scenario that serves as the story's foundation.

Passion and organization are required to come up with a compelling plot. Define your primary characters' backgrounds, demeanor, and desires after choosing a genre that speaks to you whether it be horror, suspense, romantic comedy, or education. Select a location that fits your story's tone and atmosphere, and create a brief outline of the main conflict, highlighting the obstacles and choices your protagonist faces. Create supporting characters who have their own goals, connections, and motives as well as tension and suspense through shifts of plot points. Think about the subjects and themes you wish to accentuate. Finally, describe how the problem will be resolved, making sure it is satisfying and wraps up pending concerns. As you develop your writing, be willing to make changes to your drafts and design.

TSOHPM

Exercise #6: Designing Plot

Establish Clear aims and Motivations: Begin by establishing your major characters' aims and motivations. What do they hope to accomplish, and why? Clear objectives create a firm framework for your plot and drive the actions of your characters, giving them a feeling of purpose and direction.

Create Conflict and hurdles: Include conflicts and hurdles that your characters will face on their path. Obstacles offer dimension to the narrative and keep the audience involved as they pull for the characters to overcome difficulty, whether they be internal problems, interpersonal disputes, or external challenges.

Create a powerful beginning, middle, and end: Create a plot with a clear beginning, middle, and end. The setting, characters, and starting conflict should all be established in the beginning. The middle of the story builds the plot, introduces twists, and deepens the tension, while the end resolves the main issues and delivers a satisfactory finish.

Create Character Arcs: Create intriguing character arcs by allowing your characters to grow and change throughout the plot. Characters should be confronted with obstacles that push them to confront their defects or inadequacies, resulting in personal growth and transformation.

Maintain Consistent Pacing: Keep the pacing of your story consistent to keep the audience engaged. Tension and action should be balanced with quieter, introspective times. Aim for a lively and well-paced narrative, avoiding long periods of monotony or excessive action without meaningful development.

Surprise Your Audience: To keep your audience interested, introduce unexpected twists and turns. Surprises and disclosures provide excitement and reduce predictability while retaining internal logic inside your plot. A well-crafted story should challenge readers' or viewers' assumptions and keep them wondering.

Styles and Traditions

Storycraft is the art and skill of writing narratives that takes many shapes and varieties, reflecting an extensive realm of human creativity. Storytelling encompasses ancient oral traditions as well as the storylines of film and television. The art of telling stories is steadily evolving through digital media, theater, poetry, and emerging technologies like virtual reality. Storycraft's versatility ensures its enduring capacity to fascinate and connect people across time and platforms.

Oral Tradition: The oldest type of storytelling, frequently containing myths, stories, and folktales and passed down verbally over centuries.

Written Tradition: Storytelling through written word, which includes novels, short stories, and poems.

Epic Poetry: Narrative poetry with a vast and exalted style that often honors heroic achievements.

Folklore: A community's traditional stories, practices, and beliefs, such as fairy tales, fables, and myths.

Drama and Theater: Storytelling through performance, as seen in plays and theatrical shows, with aspects such as dialogue, story, and character development.

Film and Cinema: Visual narrative using film as a medium, including cinematography, editing, and sound design.

Television Series: Serialized narrative on the tiny screen, sometimes with longer character arcs and intricate plot development.

Comic Books and Graphic Novels: Narrative storytelling using graphics and text, typically spanning genres ranging from superhero fiction to autobiographical narratives.

Digital Storytelling: the use of digital platforms such as interactive media, online tales, and multimedia storytelling.

Video Games and Virtual Reality: Storytelling where the audience has agency, making decisions that affect the narrative conclusion; interactive fiction.

Radio and Podcasts: Audio storytelling that uses voice, sound effects, and music to express storylines.

Surreal and Absurdist Fiction: Storytelling that questions traditional reality, frequently involving dreamy or surreal settings.

Historical Fiction: is defined as stories set in the past that combine fictitious elements while depicting historical events and individuals.

Magical Realism: The seamless incorporation of magical or fantastical aspects within a realistic narrative, typically exploring the extraordinary inside the commonplace.

JANGLED JESTER

Exercise #7: Explore a New Genre

Identify core aspects: Begin by determining the core aspects that define the storytelling style you've chosen. Whether the emphasis is on character development, narrative structure, or thematic components, identifying the core characteristics aids in the establishment of a framework for study.

Investigate narrative tactics: Look into the specific narrative tactics used in the style. Examine the story's presentation, including characteristics such as point of view, tempo, and language use. Understanding these strategies gives light on the storyteller's stylistic choices.

Investigate symbolism and motifs: Examine the storytelling style for repeating symbols, motifs, or themes. These motifs frequently carry deeper meanings and help to the narrative's overall cohesiveness. Analyzing symbolism improves your understanding of the story's layers.

Consider the cultural and historical background: Investigate the cultural and historical background in which the storytelling style arose. Understand how societal influences, historical events, or cultural movements affected the style's narrative conventions and themes.

Characterization: Take note of how characters are developed and portrayed in the storytelling approach. Character arcs, motivations, and connections are important in expressing the narrative and reflecting the style's distinct features.

Emotional impact: Think about the emotional impact that the storytelling style is attempting to generate. Examine how the story provokes feeling, engages the audience, and creates an interesting environment or message. Understanding the emotional components aids in understanding the intended impact on readers or viewers.

CHAPTER 2

THE PHILOSOPHER

JANGLED JESTER

Annie and Lucian navigate the winding, rugged roads left by Hurricane Helene's path as the sun peeks through the clouds, casting a soft spell of light over the Appalachians. The truck rumbles over uneven terrain, and Annie's shifts focus between the landscape and her notebook, scribbled with observations about the geological impacts of the recent storm.

"Did you notice how the soil erosion patterns changed after the hurricane?" Annie asks, tapping her pen against the notepad. "It's fascinating — and troubling. The way the mountains are layered, I'd expect certain areas to hold better, but..."

Lucian glances at her, intrigued. "You think the hurricane's intensity caught everyone off guard? The models didn't predict this level of destruction."

"Exactly," she replies, her brow furrowing. "It's almost like the mountains absorbed the storm in a way that accelerated the damage. We need to collect more data to understand why. This isn't just about rebuilding; it's about preventing this kind of catastrophe in the future."

As they crest a steep incline, more devastation unravels — rows of wrecked homes and toppled trees lining the valley. Annie's heart stops at the sight of people gathered in clusters, their faces drawn and weary.

She turns to Lucian, her voice dropping. "It's hard seeing them like this. Appalachia is joyful in song. Now it feels like a trap, cut off from the world, insulated to suffocation."

Lucian nods, his expression thoughtful. "There's a lot out there still suffering, but they're not alone. Listen to how they're helping each other. We're helping them. Bringing each other light."

Annie's thoughts shift to Plato's Allegory of the Cave as she sees a woman cradling a muddy quilt, surrounded by neighbors sharing their tools and resources. "You're right. It's like living in a cave, only seeing the shadows of devastation, believing that's all there is. They think no one cares."

Lucian watches a fellow scientist of Annie's in mucky overalls approach a small group, hesitant but determined. "Our work isn't just about studying and delivering supplies; it's about showing these people they're seen, that the outside world hasn't forgotten them."

They arrive at a church-turned-shelter, where the local pastor leads a group in prayer. Annie and her team joins out of respect, bowing their heads, while Lucian stands back, observing. When the prayer concludes, they unload supplies, offering food and clean water to the locals, who share stories of resilience amid loss.

During a break, Lucian takes a seat on a nearby log, watching Annie interact with the community. "You seem at ease here," he remarks.

She glances over, wiping sweat from her brow. "I may not share their beliefs, but I believe in their strength. They're coping, creating a network to support one another. It reminds me that while I've stepped away from certain philosophies, I'm not cut off from the core values that connect all of us."

As they continue their work throughout the day, the labor is heavy, but laughter mingles with gratitude, the metal bulldozers singing all night, and the town working together to quilt a new hope. By evening, they pile back into the truck, the sun setting over the mountains, casting a golden hue over destruction.

Lucian looks back at the valley, glimpsing both fear and fierce determination in the locals' eyes. "We're stronger than these storms we face."

Annie nods, her spirit renewed. "Yes — every time, we rise from the rain, moving toward the light beyond the shadow."

Plato's Allegory of the Cave

From "The Republic" — a debate between Plato, Socrates, and the Athenian citizens — comes Plato's allegory of the cave, a well-known philosophical metaphor. In the allegory, Plato likens people to prisoners in a cave. They do not know they are imprisoned. Their only view is a shadowy wall illuminated by a campfire. The inhabitants think the only world that is real is the shadow realm in the cave they see, but one caveman manages to escape and develops a new philosophical outlook.

He sets out into the untrodden world and discovers his worldviews are illusory. The philosopher returns to the cave and tries to set the other prisoners free, but sadly, his ideas aren't well received. They tell him he's crazy. The philosopher perseveres nonetheless as he is committed to truth and knows he has experienced another world and greater truth. He is inspired, and he thinks he can help some of the other cave inmates escape eventually, too.

Plato's allegory doesn't tell us where the philosopher journeys next, but it does establish a framework for beginning our own daring philosophical adventure. In "What is Philosophy?" Gilles Deleuze and Félix Guattari (1994) define it as the creation of concepts and the philosophical journey as a social, emotional, and transformational process that is contingent on complex concept-mapping. The concept of a butterfly, for instance, isn't only in the name. A butterfly is also in its shape, color, and biological ourpose. There are symbolic and cultural significances about butterflies in human art and language as well as anatomical associations with it in science.

Philosophy is a journey. It has many dimensions as explored by renowned thinkers like Plato, Immanuel Kant, and Bertrand Russell. Plato defines philosophy as the love of wisdom and a passionate desire for insight and knowledge. Philosophy, according to Kant, is the study of reason and limitations of human thinking. Russell defines philosophy as an attempt to organize and integrate large systems of knowledge spanning many fields, with the ultimate goal of achieving a more coherent and fundamental understanding of nature and reality. Plato, Kant, and Russell share a common emphasis on the pursuit of insight, clarity, and exploration.

In today's world, most may be unfamiliar with the meaning of philosophy or fail to see its importance in everyday living. Some may regard it as arcane or exclusive to academics and intellectuals, while others disregard it as unimportant and impractical. Philosophy is essential in that it teaches humans critical thinking skills, improves the ability to comprehend complicated subjects, and supports growth for a greater awareness and sense of being.

The Foolish Explorer

Jangled Jester's archetypal character embodies the essence of the fool — a mix of ignorance and bravery. He stands at the mountain's edge, blissfully unaware of the dangers lurking around him: a sudden gust of wind could topple him over, or a mountain lion might be hiding just beyond the bushes. Yet, he stops to smell the flowers and continues his journey, undeterred by the risks. This foolish explorer revels in the day's opportunities, embracing self-esteem while dancing on the edge of uncertainty.

Another core principle of Jangled Jester's philosophy is the idea of being the foolish explorer in the absurd. It encourages us to take risks, put ourselves out there, and adopt the mindset of a philosopher. Think for yourself. Question the shadows that loom in the corners of your mind. Escape the cave of ignorance. Climb the mountain of knowledge, equipped with a bag of rocks — just in case a dramatic mountain lion crosses your path, and a notebook to document and tell the story of it later. Nature's landscapes are rife with tumultuous conflicts, echoing wars, tension, and plays. These elements are the rising actions, peaks, and valleys in the natural stories we cherish.

The philosopher embarks on an endless journey of curious exploration. In traditional tarot, the fool's character corresponds to the number 0, symbolizing infinite potential. You needn't view tarot mystically to appreciate the psychological depth of starting from zero.

There is no true beginning or end in philosophy; there is only transformation, a continuous cycle where inanimate nature becomes a canvas for reflection. The philosopher is the embodiment of conscious awareness and creative contemplation. This force invokes curiosity that propels rockets into space, composes poems and symphonies, and embraces the full spectrum of human emotion — passion, grief, joy, and wonder.

In our daily lives, the role of the philosopher can manifest in simple yet profound ways. It invites us to be curious, to question the status quo, and to find beauty in things. As we navigate our own mountains, we can ask ourselves: What else is there to contemplate? What truths lie hidden beneath the surface of our experiences? Each day presents a new opportunity to explore, to learn, and to grow. The foolish explorer reminds us that the journey is just as important as the destination. So grab your bag of rocks and journal, take a deep breath, and step boldly into the unknown.

Exercise #8: Tips for Designing a Philosophy

Philosophy gives us the opportunity to confront ourselves and moral quandaries, make educated judgments, and traverse life's difficulties with greater purpose and clearer perception. To fully grasp and benefit from philosophy, remember to apply all these steps when building your philosophy:

Read. Reading philosophical texts, both classic and contemporary, provides a rich source of ideas and inspiration. You may start with classics like Plato, Aristotle, and Descartes, or read works by contemporary philosophers like Judith Butler and Albert Camus.

Discuss. Engage in philosophical discussion to develop your thinking skills and clarify your ideas. Join a philosophy club or discussion group online or have conversations with friends and family to explore philosophical concepts and perspectives.

Identify. Take time to identify your core values, like honesty, fairness, and compassion. Write them out in logical statements. Your values will serve as a template for decision-making, goal setting, and behavior modification.

Examine. Consider ethical dilemmas and think about ways you would respond or react to them. This will help you develop reasoning skills and strengthen your ethics and values that may change over time or be affected by unknown factors.

Connect. Break complex subjects and systems down into digestible bits and look for underlying concepts that connect them when constructing a philosophy, aiming for clarity and coherence in your speech to create shared understanding.

Apply. Don't be afraid to consult and apply the philosophies of other thinkers. For example, if you are struggling with an ethical dilemma, you may find guidance in the works of a treasured philosopher.

Reflect. Reflect on your experiences and thoughts. What is important to you? What patterns do you see in yourself, others, and nature? What's worth changing, keeping, and what will you do to maintain it?

Ask Fertile Questions. Asking fruitful questions will lead to new, useful questions. Initially developing a great question can be tough, so some keys to brainstorming questions are: being curious about how things work, having a desire to listen, learn and observe, challenging assumptions and the status quo, and asking follow-up questions.

Examples of Philosophical Questions

What is the nature of reality?

Is reality objective, or subjective?

How can we know anything for certain?

What is the nature of knowledge?

What is the meaning of life?

What is our purpose as human beings?

Is there a moral basis for human behavior?

What are the ethical standards that should guide our morals and actions?

What is the relationship between the mind and body? Is the mind separate from the body?

What is the nature of consciousness? Do non-human beings have consciousness?

What is the role of language in shaping our perceptions and understanding of the world?

What is the role of art and aesthetics in human life?

How should we deal with conflicting / contrasting values, actions, and beliefs?

Is there a universal code of ethics that all cultures should adhere to?

Is free will an illusion, or is it a real phenomenon?

What is the nature of time? Does time exist independently of human perception?

What is the nature of truth? Is there a single, objective truth, or are there multiple truths?

How should we balance individual rights and common good?

What is the relationship between faith and reason? Can they be reconciled?

Philosophers to Explore

Ayn Rand (Objectivism): Rand's philosophy of Objectivism emphasizes reason, individualism, and the pursuit of objective knowledge. These align with physicalist magick's focus on the form and function of transformation grounded in reality. Her rejection of mysticism and emphasis on rational self-interest provide a foundation for secular occultism as a methodical exploration of the world.

Albert Camus (Existentialism and Absurdism): Camus' exploration of the absurd—the conflict between human desire for meaning and the universe's indifference—parallels the existential uncertainty often embraced in secular occultism. His call to live authentically despite this tension resonates with physicalist magick's transformative power to craft meaning and purpose.

Karl Popper (Critical Rationalism): Popper's approach to science, emphasizing falsifiability and the ongoing critique of theories, can enhance the practice of physicalist magick by promoting a rigorous, empirical method of engaging with reality, thus supporting non-dogmatic, scientific inquiry.

Plato (Idealism and Metaphysics): While Plato's theory of forms contrasts with physicalist magick's focus on material transformation, his allegory of the cave can inspire secular occultists to seek deeper understanding of the "shadows" of reality. Exploring his dialectical method can enhance knowledge-seeking curiosity.

Aristotle (Empiricism and Teleology): Aristotle's emphasis on observing the natural world and understanding causes aligns with the pragmatic aspects of physicalist magick. His concept of teleology (purpose or end-goal) connects to the design of transformation, while his empirical approach supports secular inquiry.

Friedrich Nietzsche (Will to Power and Nihilism): Nietzsche's idea of the will to power echoes the transformative essence of physicalist magick as an act of creative self-overcoming. His critique of traditional morality and metaphysics complements the secular occultist's rejection of dogma and embrace of curiosity.

Jean-Paul Sartre (Existentialism and Freedom): Sartre's focus on radical freedom and the responsibility to create one's essence aligns with the self-directed transformation of physicalist magick. His concept of "bad faith" warns against self-deception, a critical reminder for those seeking knowledge through secular occultism.

Baruch Spinoza (Pantheism and Determinism): Spinoza's identification of God with nature provides a naturalistic perspective that resonates with physicalist magick's grounding in the material world. His deterministic view encourages understanding the interconnectedness of things, vital for secular curiosity.

Ludwig Feuerbach (Humanism and Materialism): Feuerbach's critique of religion and emphasis on human essence as material and natural supports a secular, human-centered approach in physicalist magick, focusing on human potential for self-realization.

Lucretius (Epicureanism and Materialism): Lucretius' poem De Rerum Natura celebrates the materialist worldview, offering a poetic exploration of atoms, the void, and natural phenomena. His ideas inspire physicalist magick by framing transformation as part of the natural order, while his curiosity fuels secular inquiry.

Bertrand Russell (Logical Positivism and Skepticism): Russell's advocacy for clear, logical thinking and his skepticism towards traditional metaphysics can guide practitioners of physicalist magick towards a more empirical, evidence-based practice, questioning supernatural explanations in favor of natural ones.

Thomas Nagel (Philosophy of Mind): Nagel's exploration of consciousness and his critique of reductive materialism can provide a nuanced perspective for physicalist magick, where understanding mental phenomena is crucial for personal transformation while still acknowledging the limits of physical explanations.

Who else can you add to this list?

My Beloved Philosophers

_____

_____

_____

_____

_____

_____

_____

_____

_____

Exercise #9: A Deeper Dive into a Thinker's World

Choose a philosopher: To begin, select a philosopher whose ideas and writings interest you. Choose a philosopher whose work resonates with your interests or challenges your perspectives, whether they are well-known or less well-known.

Read the philosopher's primary texts: Immerse yourself in the philosopher's primary texts. To obtain a personal grasp of their beliefs, read their significant works. Take notes, mark important portions, and consider the basic themes given in their texts.

Investigate secondary sources: Supplement your reading with credible secondary sources. Examine scholarly papers, analyses, and commentaries on the philosopher's work. This will bring more insights, perspectives, and historical context to your understanding.

Trace influences: Look into the philosopher's intellectual lineage. Identify thinkers who affected them and those who were influenced by their beliefs. This contextual investigation assists you in situating the philosopher within a broader philosophical environment.

Consider the philosopher's historical and cultural context: Understand the philosopher's historical and cultural context. Investigate their time's cultural, political, and philosophical currents, as these variables frequently affect and inform the philosopher's perspectives.

Reflect and respond: As you read, engage in reflection tasks. Think about how the philosopher's ideas connect to your own beliefs and experiences. Make a list of your ideas, questions, and criticisms. This procedure not only broadens your awareness but also promotes your intellectual development.

Participate in debates and communities: Take part in online debates and communities related to the philosopher's work. Engaging in discourse with people who are interested in the same philosopher allows you to exchange ideas, get various views, and deepen your understanding.

Glossary of Philosophical Concepts

Absolutism: The belief in the existence of absolute truths, principles, or moral standards.

Abstraction: A cognitive and artistic process of simplifying or generalizing complex ideas, objects, or concepts to represent them in a more conceptual, non-literal, or symbolic form, often used to explore fundamental characteristics or patterns.

Absurdism: A philosophical perspective that explores the inherent absurdity and meaninglessness of life.

Aesthetics: The study of beauty, art, and the principles that govern our sense of what is aesthetically pleasing.

Agnosticism: The belief that knowledge about certain topics, particularly the existence of gods or ultimate realities, is inherently uncertain or unknowable.

Animism: The belief that natural objects, phenomena, and animals possess spiritual or supernatural qualities and consciousness.

Anthropomorphism: The attribution of human characteristics, emotions, or behaviors to non-human entities, such as animals, objects, or deities.

Atheism: The absence of belief in the existence of supremely sacred deities.

66

Behaviorism: A psychological and philosophical approach that focuses on observable behavior as the primary subject of study, emphasizing the role of conditioning and environmental factors in shaping human and animal behavior.

Capitalism: An economic and political system characterized by private ownership of the means of production, market competition, and the pursuit of profit as a driving force for economic activity.

Categorical Imperative: An ethical principle formulated by Immanuel Kant that emphasizes acting on moral rules universally applicable to all.

Collectivism: A social and political ideology that emphasizes the importance of collective or group interests and cooperation over individual autonomy and competition.

Communism: A socio-economic and political ideology advocating for the collective ownership of the means of production, the absence of social classes, and the equitable distribution of resources.

Consequentialism: An ethical theory that evaluates actions based on their consequences, seeking the greatest overall good.

Cogito, Ergo Sum: Latin for "I think, therefore I am," a famous statement by Descartes reflecting the certainty of self-awareness.

Cynicism: A skeptical or pessimistic attitude towards the motives and integrity of others, often characterized by distrust, sarcasm, or a belief in self-interest as the primary motivation for human actions.

Darwinism: A scientific theory developed by Charles Darwin that explains the process of evolution by natural selection, where species adapt to their environments over time through the survival and reproduction of traits that offer advantages for survival.

Dasein: A concept in existentialism referring to the unique, subjective experience of being.

Deduction: A method of reasoning from general principles to specific conclusions.

Determinism: The belief that events are determined by previous causes, implying that free will may be an illusion.

Deontology: An ethical theory that emphasizes the inherent rightness or wrongness of actions, regardless of their consequences.

Dogma: A set of established beliefs or doctrines, often religious or ideological, that are considered authoritative and not subject to questioning or doubt within a particular belief system or organization.

Dualism: The view that the mind and body are distinct substances, often associated with René Descartes.

Empiricism: The belief that knowledge is primarily derived from sensory experience and observation.

Epistemology: The branch of philosophy that explores the nature, sources, and limits of knowledge.

Epicureanism: A philosophy that promotes a simple and moderate life focused on pleasure and the avoidance of pain.

Ethics: The study of moral principles, values, and what is considered right or wrong behavior.

Eudaimonia: In virtue ethics, the state of flourishing, often associated with living a virtuous life.

Existentialism: A philosophical movement that emphasizes individual freedom, choice, and the meaning of existence.

Falsifiability: The quality of a statement, hypothesis, or theory that can be tested and potentially proven false through empirical evidence or observation, a key criterion in scientific inquiry.

Freudianism: A psychological and psychoanalytic theory developed by Sigmund Freud that explores the role of the unconscious mind, repressed desires, and early childhood experiences in shaping human behavior and personality.

Hedonism: The pursuit of pleasure as the highest good or the ultimate goal of life.

Hermeneutics: The study of interpretation, particularly of texts, and understanding their deeper meanings.

Holism: A philosophical and scientific approach that emphasizes the interconnectedness and unity of systems, where the whole is greater than the sum of its parts, often applied in various fields, including ecology, medicine, and philosophy.

Idealism: The belief that reality is fundamentally mental in nature.

Illusionism: An artistic and philosophical approach that seeks to create the illusion of reality or three-dimensionality in two-dimensional artworks, often through techniques of perspective and meticulous detail.

Immortalism: A philosophical and ethical belief that advocates for the pursuit of immortality or significantly extended human lifespan, often through scientific or technological means.

Individualism: A philosophical and social ideology that emphasizes the importance of individual freedom, autonomy, and self-reliance, often valuing the rights and interests of the individual over collective or societal values.

Induction: A method of reasoning from specific observations to general conclusions.

Liberalism: A political and philosophical ideology that promotes individual rights, democracy, limited government intervention in personal

70

and economic affairs, and the rule of law as essential principles for a just and free society.

Logic: The study of reasoning, argumentation, and the principles of valid inference.

Magick: The design of function; processes of change.

Materialism: The belief that reality consists only of physical matter and that mental phenomena can be explained by physical processes.

Metaphysics: The study of fundamental questions about reality, existence, and the nature of the universe.

Moral Relativism: The belief that moral values are not universally applicable but are dependent on cultural or individual contexts.

Mysticism: A spiritual and philosophical belief or practice that seeks direct, personal experience and union with a transcendent or divine reality through contemplation, meditation, or mystical experiences.

Naturalism: A philosophical and literary movement that emphasizes the role of natural processes and empirical observation in explaining and understanding the world, often portraying human behavior as determined by natural forces.

Nihilism: The belief that life lacks inherent meaning or value.

Objectivism: A philosophical system developed by Ayn Rand that emphasizes objective reality, reason, individualism, and rational self-interest as fundamental principles for human existence and morality.

Ontology: The study of the nature of being and existence.

Panpsychism: A philosophical theory that suggests that consciousness or mind is a fundamental property of the universe, and that all entities, not just humans or animals, possess some form of consciousness.

Pantheism: The belief that the divine or God is identical to the universe or nature.

Phenomenology: A philosophical approach that seeks to describe and understand subjective human experiences.

Phenomenon: The appearance or manifestation of an object as perceived by the senses.

Philosophy: The systematic study of fundamental questions concerning existence, knowledge, values, reason, mind, and language, often involving critical thinking, rational inquiry, and a search for understanding the nature of reality and human experience.

Physicalism: The philosophical view that everything in the universe, including mental states and consciousness, is ultimately physical or can be explained by physical laws and entities, with no independent non-physical substances.

72

Pluralism: The belief that there are multiple perspectives, values, or truths that can coexist.

Postmodernism: A philosophical, cultural, and artistic movement that emerged in the mid-20th century, challenging traditional notions of truth, authority, and meaning, often characterized by a skepticism of grand narratives, an emphasis on cultural diversity, and a focus on subjective interpretations.

Pragmatism: A philosophy that emphasizes practical consequences and utility as criteria for truth and meaning.

Probabilism: A philosophical and ethical approach that suggests making choices based on the most probable or likely outcomes, especially in situations where moral or practical certainty is not attainable.

Qualia: The subjective, conscious experiences or qualities of sensory perceptions, emotions, or mental states, such as the specific sensation of seeing the color red or experiencing the taste of chocolate, which are difficult to fully convey or explain to others.

Rationalism: The belief that reason is the primary source of knowledge and that it can lead to certain, foundational truths.

Realism: A philosophical and artistic movement that emphasizes representing the world as it is, often with a focus on the everyday and the objective, without idealization or embellishment.

Reductio ad Absurdum: A logical argument that demonstrates the absurdity of a claim by taking it to its extreme.

Relativism: The view that truth, morality, or meaning is relative and depends on individual or cultural perspectives.

Romanticism: A cultural, artistic, and literary movement that emerged in the late 18th and early 19th centuries, emphasizing emotions, imagination, nature, individualism, and the expression of personal experiences and sentiments.

Secularism: A philosophical and social stance that champions the freedom of individuals to express and practice their religious or non-religious beliefs without discrimination, ensuring that both religious and secular perspectives are equally respected in public discourse and affairs.

Skepticism: The view that true knowledge is difficult to attain, often leading to doubt or suspension of judgment.

Socialism: A socio-economic and political ideology that advocates for collective or state ownership of the means of production, equitable distribution of resources, and the reduction of economic inequality through government intervention and social programs.

Solipsism: The belief that one's own mind is the only reality, and external reality may not exist.

Stoicism: A philosophy that advocates self-control, rationality, and acceptance of fate as a path to virtue and tranquility.

Symbolism: A literary, artistic, or cultural movement that employs symbols, signs, or objects to represent deeper meanings, ideas, or concepts beyond their literal interpretation, often conveying complex emotions or themes.

Tabula Rasa: Latin for "blank slate," a concept suggesting that individuals are born with a clean mental slate and acquire knowledge through experience.

Teleology: The study of purpose or the belief that natural phenomena are directed toward specific goals.

Theism: Belief in the existence of a deity or deities.

Totalitarianism: A political system characterized by centralized and absolute control over all aspects of public and private life, often by a single ruling party or leader, with limited individual freedoms and often characterized by censorship and surveillance.

Transcendentalism: A philosophical and literary movement that emerged in the 19th century, emphasizing the inherent goodness of people and nature, the importance of self-reliance, and a belief in the transcendental or spiritual dimension of human existence.

Utilitarianism: A moral theory that promotes actions that maximize overall happiness or utility.

Utopianism: A philosophical and social belief in the possibility of creating an ideal, perfect, and harmonious society or world characterized by social and political perfection, often through visionary or idealistic means.

Virtue Ethics: An ethical approach that emphasizes the development of virtuous character traits.

Fallacies

Fallacies are logical mistakes that can damage the quality and validity of an argument. They frequently exist in various forms of discourse, ranging from casual talks to intellectual arguments. Recognizing and understanding fallacies is essential in philosophy and critical thinking because it allows people to detect erroneous arguments and engage in more rational and constructive debates.

Ad Hominem: Attacking the person making the argument instead of addressing the argument itself.

Ad Misericordiam (Appeal to Pity): Appealing to sympathy or pity instead of providing valid reasons for a conclusion.

Ad Populum (Bandwagon Fallacy): Arguing that something must be true or good because it's popular or widely accepted.

Ad Verecundiam (Appeal to Inappropriate Authority): Citing an authority figure in an unrelated field as an expert.

Ambiguity: Using language that is intentionally vague or unclear to create confusion or evade the burden of proof.

Anecdotal Evidence: Using personal anecdotes or isolated examples as evidence for a general claim.

Appeal to Authority: Relying on the opinion of an authority figure instead of providing evidence or sound reasoning.

Appeal to Ignorance: Arguing that something is true because it hasn't been proven false or false because it hasn't been proven true.

Begging the Question (Circular Reasoning): Assuming the truth of the conclusion in the premises.

Burden of Proof: Shifting the responsibility to prove or disprove a claim onto the wrong party.

Circular Reasoning: Using a claim to support itself without providing any real evidence.

Composition: Assuming that what is true of the parts must also be true of the whole.

Division: Assuming that what is true of the whole must also be true of its parts.

Equivocation: Using ambiguous language or terms with multiple meanings to mislead or confuse.

False Dichotomy: Presenting only two extreme options as if they are the only possibilities, ignoring middle ground or alternatives.

Genetic Fallacy: Dismissing an argument or claim based on its origin or source rather than its merit.

Hasty Generalization: Drawing a broad conclusion based on insufficient or unrepresentative evidence.

78

No True Scotsman: Reinterpreting evidence or redefining terms to exclude counterexamples.

Post Hoc (False Cause): Assuming that because one event happened after another, the first event caused the second.

Red Herring: Diverting attention from the main issue by introducing unrelated or irrelevant information.

Slippery Slope: Claiming that one event will inevitably lead to a chain of negative events without providing sufficient evidence.

Straw Man: Misrepresenting or distorting an opponent's argument to make it easier to refute.

Tu Quoque (Appeal to Hypocrisy): Dismissing an argument by pointing out the hypocrisy of the person making the argument.

Rules of Logic

Inference: In formal logic, rules of inference are essential ideas or guidelines used to create acceptable arguments and draw logical conclusions from premises. They provide a methodical and dependable approach of reasoning about assertions and propositions. Here are some key principles linked to inference rules:

Validity: An argument is valid if, given that the premises are true, the conclusion must also be true. Rules of inference help ensure the validity of an argument.

Soundness: An argument is sound if it is both valid and all of its premises are true. Sound arguments are considered reliable and persuasive.

Logical Connectives: Rules of inference often involve logical connectives such as AND, OR, NOT, IF-THEN, and their corresponding rules for combining propositions.

Proofs

A logical proof is a step-by-step demonstration of a conclusion's validity or truth based on a set of premises and the application of inference rules. A logic proof's purpose is to produce a clear and rigorous argument that demonstrates how the conclusion logically follows from the given premises. Here are a few important points to remember about logic proofs:

Premises: A logic proof starts with a set of premises, which are the statements or propositions assumed to be true.

Conclusion: The proof aims to establish the truth of a specific conclusion based on the given premises.

Rules of Inference: During a proof, rules of inference are applied systematically to derive intermediate conclusions or steps. These rules ensure that each step in the proof is logically valid.

Derivation: Each step in the proof is derived from previous steps using valid rules of inference. The process continues until the conclusion is reached.

Justification: In a logic proof, each step must be justified and explicitly stated. This typically involves citing the rule of inference applied and referencing the premises or previously derived steps.

Completeness: A complete logic proof demonstrates that the conclusion follows logically and necessarily from the premises. It leaves no room for doubt or ambiguity.

Practical Reasoning

In everyday life, practical reasoning entails making judgments, solving issues, and responding. It frequently involves a blend of critical thinking, judgment, and ethical factors. While there aren't as many tools for practical reasoning as there are for formal logic, there are various cognitive and practical tools and techniques that can help you make solid and informed decisions. Here are some practical reasoning tools:

Decision-Making Models: Various decision-making models, such as the Rational Decision-Making Model and the Bounded Rationality Model, provide structured approaches for making choices based on available information and preferences.

SWOT Analysis: A strategic planning tool that assesses an organization's Strengths, Weaknesses, Opportunities, and Threats to guide decision-making and strategy development.

Cost-Benefit Analysis: A systematic evaluation of the pros and cons of different options, where the benefits and costs are quantified and compared to inform decisions.

Ethical Frameworks: Ethical theories and frameworks like utilitarianism, deontology, and virtue ethics can help individuals assess the moral implications of their decisions.

Problem-Solving Methods: Techniques like root cause analysis, brainstorming, and the scientific method are employed to identify and solve problems effectively.

Risk Analysis: Tools such as risk matrices and decision trees can help assess the likelihood and impact of risks associated with different choices.

Prioritization Matrices: A tool for ranking and prioritizing options or tasks based on criteria such as importance, urgency, or impact.

Scenario Planning: A method for exploring various possible future scenarios and their implications to make more robust decisions.

Critical Thinking: Developing critical thinking skills, including analyzing information, questioning assumptions, and evaluating evidence, is essential for practical reasoning.

Intuition and Gut Feeling: Sometimes, relying on intuition and instincts, backed by experience, can be a valuable tool for making quick decisions.

Consultation and Collaboration: Seeking advice from experts or consulting with colleagues and stakeholders can provide diverse perspectives and insights.

Data and Information Gathering: Collecting relevant data, conducting research, and staying informed are essential for making informed decisions.

Goal Setting: Clearly defining objectives and desired outcomes helps align decisions with long-term goals.

Feedback and Evaluation: Regularly assessing the outcomes of decisions and adjusting course as necessary is crucial for continuous improvement.

Time Management Techniques: Effective time management tools and strategies help individuals allocate their time efficiently when considering multiple tasks or options.

Reflection and Mindfulness: Taking time to reflect on one's values, goals, and priorities can enhance decision-making clarity.

Communication Skills: Effective communication and negotiation skills are essential for reaching agreements and making collaborative decisions.

Stakeholder Analysis: Identifying and understanding the interests and concerns of stakeholders can guide decisions that consider a broader impact.

Emotional Intelligence: Recognizing and managing emotions, both one's own and others', can be important in decision-making processes.

Feedback Loops: Establishing mechanisms for receiving and responding to feedback on decisions can lead to more adaptive and effective decision-making.

JANGLED JESTER

Exercise #10: Dedicate a Philosophy Journal

Grab a notepad or launch a digital document to start journaling. Start by jotting down the following inquiries, then take your time to thoughtfully respond to each one:

What do you think the nature of existence and reality is?

What are your fundamental beliefs and values?

What are morals and ethics, in your opinion?

According to your beliefs, what is the point of human life?

Do you hold any metaphysical or spiritual beliefs?

What do you think about how people interact with the natural world?

What influence does logic, feeling, or intuition have on your choices?

CHAPTER 3

THE PHYSICALIST

Annie and Lucian drove cautiously along the mountain roads, their truck bed rattling with solar panels and tools. The work wasn't glamorous, but it was urgent. Every stop brought them to homes where power outages stretched into days, and the chill of fall nights was settling in.

At their first house, the McAllister family greeted them with grateful smiles and instant-mix coffee heated over a campfire. Annie thought it was some of the best she'd ever had, knowing the care they'd taken purifying the water. While Lucian and the team climbed onto the roof to install panels, Annie stayed on the porch, chatting with Ruth McAllister, the eldest daughter, a high school senior poring over textbooks by candlelight.

"Y'ever hear of Nikola Tesla?" Annie asked, glancing at Ruth's math notes. "He said numbers like three, six, and nine are the key to the universe."

"Numbers aren't magic," Ruth replied, her gaze dropping to her notes. "They're tools for describing patterns, like how electricity flows or how storms form."

Annie shrugged and grinned. "You're telling me."

Ruth continued reading, her expression thoughtful.

"So," Annie went on, "if numbers are just tools, why do they keep popping up in nature? Triangles in circuits, spirals in storms? Feels like there's something deeper, something alive in all of it."

Ruth considered this and finally responded, "Patterns don't make something alive. But maybe what feels alive is how it all connects. Circuits, weather systems, even us — it's all matter, moving and reshaping."

Their conversation was interrupted by Ruth's father, Jack McAllister, who stepped up onto the porch. He wiped his hands on his jeans, nodding toward the roof.

"Storms will bust this solar setup," he said, gesturing to a pile of pine debris scattered in the yard. "Good for sunny days, useless when the clouds roll in."

Annie shook her head, ready with a response. "Stormy weather does reduce the direct efficiency of solar panels, but it doesn't make them useless. Panels can still generate electricity from diffuse light, even on cloudy days. Plus, the future of renewable energy isn't about relying on just one source.

"We're working on hybrid solar-hydro systems, for instance," she explained. "They store energy during sunny periods by pumping water to a reservoir, then use that water to generate electricity when it's dark or stormy. Floating solar panels are another idea—they sit on reservoirs, reduce evaporation, and still capture energy.

"The real challenge is storage and integration, and technologies like grid-scale batteries and pumped hydro storage are tackling that. Storms like these remind us why decentralized systems — local solar combined with storage — are essential. It's not about perfect weather; it's about building systems that adapt and thrive no matter what."

Jack nodded slowly, his skepticism softening. "Makes sense. Guess it's not just the panels we need, but a whole way of thinking different."

"That's right," Annie said with a smile. "Ruth, why don't you come help me, my team, and your Pa here finish getting these panels installed? I think you'll enjoy it."

—

Later, over a jar of blackberry moonshine and the warm glow of Lucian's cabin lanterns, the two found themselves in a heated discussion.

"Electricity's not just wires and switches, Annie," Lucian said, leaning back in his chair. His voice had the slow cadence of someone feeling out the edges of his own thoughts. "It's alive somehow. Wild, like the storms that fling it at us. Ain't just a thing you can measure with a gauge."

Annie tilted her head, intrigued but skeptical. "Wild, sure. But alive? It's not some spirit out to get us, Lucian. It's electrons following rules, plain and simple."

Lucian took a thoughtful sip of his moonshine, his brow furrowed. "Maybe it's not alive like you or me. But it's got this... unpredictability to it. A force that feels bigger than just the sum of its parts. Like it's got a mind to it, even if it ain't a mind we can know."

Annie leaned forward, her hands clung to the jar. "You're describing emergent behavior — how complex systems can look alive when really, they're just following patterns. Electricity, storms, even ecosystems. They're all physics in motion. The unpredictability's just us not having the right tools yet."

Lucian looked at her, his expression halfway between admiration and exasperation. "So you're saying all this stuff — what folks might call mystic, maybe even magic — is just science we're too dumb to explain?"

"Exactly," Annie said. "It's not about dismissing it as magic or denying its mystery. It's about asking better questions and figuring out what makes it tick. What you're calling mystic might just be physics we haven't figured out yet.

The more we apply the scientific method, the more these patterns start to make sense. What once seemed impossible begins to reveal itself, even if the reasoning shifts along the way. You might think it's hopeless, that we can't get through weather like this if it comes back. But I'll be damned if I'm not going to try."

Physicalism

Physicalism is a naturalistic and scientific worldview that concepts like love, consciousness, and free will can be explained in physical terms and analyzed empirically. The philosophy emphasizes empirical data and reasoned inquiry to better understand existence, and it strives for a comprehensive yet developing description of reality that reduces all existence to physical attributes but leaves room for scientists to realize new physical features. Physicalism is based on a number of ideas, including materialism, reductionism, and causal determinism. One of the cornerstones of physicalism, reductivism, advocates disassembling complicated systems into simpler, lower-level parts. This is the basis of first-principles thinking.

Another essential component of physicalism is materialism, which holds that everything in the universe is made of physical matter and is governed by the rules of physics. This suggests that the physical universe is made up of various physical substances and elements. Every event has a physical cause, according to the idea of causal determinism, which is often connected to physicalism. By examining these central ideas, physicalism can make it easier to understand life and the universe.

In southern Appalachia, practicality is often rooted in first-principles thinking — a deep understanding of the physical world and the resources it provides. Families in the mountains traditionally rely on hands-on skills like blacksmithing, quilting, and woodworking, all of

94

which require an intimate knowledge of materials, processes, and the natural environment. From raising livestock to growing food, the region's people apply a grounded, physics-based approach to every task, understanding how systems work, whether it's crop rotation to enrich the soil or using renewable energy like hydroelectric power to conserve resources. These methods are based on fundamental principles that govern nature, ensuring both sustainability and efficiency.

Tillage practices are another example of how Appalachians apply physics to maintain soil integrity, as they understand the balance between minimizing disturbance and enriching the land. Techniques like canning, pickling, and smoking preserve food in ways that echo scientific principles, where each method is a way to slow down decay and extend the lifespan of natural resources. All of these actions reflect a profound respect for the environment, seen through the lens of physicalism, where the physical world is paramount, and everything is understood in terms of observable, natural processes.

Appalachians, with their deep connection to the land, have long embraced a worldview rooted in practical knowledge that aligns well with the principles of physicalism, which emphasizes the role of matter and energy in shaping our world. This lifestyle not only embodies first-principles thinking but also shows how a deep understanding of the physical world can lead to practical, sustainable living.

Physicalism aligns naturally with Southern Appalachian practicality, where the connection to the land is essential and the value of hands-on, real-world work is deeply ingrained. Rather than viewing the material world as "dirty" or "sinful," the culture in the region embraces it, seeing the land and its resources as tools for self-sufficiency and resilience. This philosophical perspective fits seamlessly into the Appalachian tradition of craftsmanship — from blacksmithing to gardening to preserving food. The act of working with the land, understanding it through first-principles thinking, and utilizing its resources responsibly reflects the heart of physicalism, where the tangible world is not just acknowledged but cherished.

Much like the physicalist view that the universe is made of matter and energy, Southern Appalachian practices reflect the idea that physical work and craftsmanship are the building blocks of a meaningful existence. In this culture, "form" and "function" are not abstract concepts; they are the practical steps that farmers take when rotating crops, the meticulous care given to livestock, and the hands-on work done to preserve food for the winter months. In this way, physicalism offers a worldview that resonates deeply with the Southern Appalachian way of life — one that prioritizes the material world as both a source of sustenance and a medium for creativity and survival.

Principles of Physics

It's normal to lack an extensive knowledge of physics, quantum mechanics, or sophisticated chemistry. These subjects form the basis of information, however, that might lead you down exciting, new, intellectual pathways even if you haven't studied them thoroughly. While physics may appear to be an abstract subject, it is simply a study that investigates and predicts the intricate chemical and transformative properties supporting our universe.

Physics can be looked at as the study of characteristics of existence that evolve with new understanding rather than a collection of fixed, immutable laws. Existence has this amazing capacity to transform yet imparts an assembly index and record of change within it. This amazing stability and equilibrium yet diversity and drama is driven by physical interactions, parameters, and principles that serve as universal foundations.

From the minute subatomic particles that make up atoms to the huge stars of galaxies, existence is made of pieces, parts, and patterns that all interact and change together. Newton's Laws define the link between motion and the forces acting on objects in classical mechanics. Quantum mechanics investigates the complex motions of tiny particles, uncertainty, and probability in nature. In essence, physical principles and physics as a discipline studies the universal choreographers, which trends to reveal a story of chemical power, abstract performance, and physical transformation.

Newton's laws describe how objects react to forces. The first law asserts that unless acted upon by a net external force, an object will stay at rest or in uniform motion. The second law describes the relationship between force, mass, and acceleration ($F = ma$), while the third law stipulates that every action has an equal and opposite reaction.

Quantum mechanics, a pillar of physics, explores a framework for understanding the actions of particles at subatomic (very small) scales. This theory introduces game-changing concepts like wave-particle duality, the dual nature of particles, electrons and photons, and such. The term "quantum" is derived from Latin and means "how much" or "how great."

The study of energy, heat, and work in physical systems is known as thermodynamics. It includes principles such as energy conservation, the growth of disorder (entropy), and the behavior of matter at different temperatures. Thermodynamics also investigates the operation of engines and the efficiency of energy conversion. It is a fundamental physics idea with applications in many scientific and engineering domains.

In the past, scientists assumed that time passed at a consistent rate for everyone, much like the tick of a clock, which made it easier to understand and calculate physical events. However, this idea has evolved significantly with the advent of Einstein's theory of relativity. According to relativity, time is not a fixed, universal constant. Instead, it can change depending on factors like speed and gravity. For example, time can move faster or slower based on your relative motion or the gravitational field you are in.

98

This view of time has been expanded by modern thinkers such as Lee Cronin and Sara Imari, who argue that time is not just an abstract concept but a tangible aspect of the physical world. They suggest that the interactions between particles and complex systems lead to what we experience as the passage of time. Time, in this context, is seen as a dynamic, physical phenomenon shaped by the forces and movements within the universe.

According to Heisenberg's Uncertainty Principle, it is impossible to know a subatomic particle's exact position and momentum at the same time. According to the concept, the more precisely you determine one of these traits, the less precisely you can know the other.

Electromagnetism is the study of electric and magnetic fields in physics. Maxwell's Equations are a collection of laws that explain how these fields interact with one another. It's like having a manual for how electric and magnetic objects work.

Exercise #11: Seeing Physical Laws that Shape Your Environment

Start by allocating time for focused observation. Choose a specific location, such as your home, a park, or a city street. Pay special attention to the physical components in your environment, such as objects, motion, light, and sound. Take note of anything that catches your eye.

While observing your surroundings, take note of the physical principles at work. Consider Newton's Laws of Motion, light and sound wave behavior, or even basic changes. Consider how objects move, the influence of gravity, or the interaction of forces.

Take pictures of any unusual phenomenon you come across. This could be the way a pendulum swings, shadows interact with light, or sound waves produced by diverse sources. Capture these moments with written notes, sketches, or even photographs to study up on later.

Reflect on your observations and the physical principles you discovered. Consider how these concepts influence your surroundings and daily life and what else you want to learn about them. Think of ways comprehending these concepts can help you appreciate the physical world more.

Physicalist Concepts

Physicalist concepts derive from a rigorous process based on empirical measurement and observation. The physical universe, regulated by natural laws and constituted of matter, is the underlying reality, according to this philosophical stance. It all starts with empirical observation, which is based on sensory experiences and scientific methodologies.

The physical world is all that exists, and everything, including the mind, supernatural realms, and subjective experience, can be explained in physical terms.

Metaphysical entities, such as God or the soul, can be explained in physical terms. The physical world includes supernatural and virtual worlds.

All causation is physical causation.

Reductionism is a valid approach to understanding complex phenomena, and everything can be explained in terms of constituent physical parts.

The physical world is causally closed, meaning all physical events are caused by other physical events.

The laws of physics govern all existing events.

Physical objects and processes have objective properties that are measurable.

There is no mind-body dualism; mental states are not separate from physical states.

There is no need to postulate any non-physical entities or properties to explain the world.

The physical world is ontologically independent of the mind or any other non-physical entity.

The physical world is self-sufficient and self-contained.

The universe is deterministic, meaning that all events are determined by prior physical causes.

The physical world is the ultimate reality, and all other domains of knowledge, such as ethics and aesthetics, are ultimately reducible to physical phenomena.

History of Physicalism

The parent of physicalism, materialism, holds that everything in the universe can be reduced to material or physical substances. It can be traced back to Democritus of Abdera (460 BCE) and his work with atomic theory. Ancient Greek philosopher, Thales (620 BCE), believed water is the fundamental substance of reality. The Roman poet Lucretius (99 BCE) was a strong materialist fond of nature, woods, open spaces, and science.

Materialism, at its heart, provides a framework for describing the bones of existence by relying on rules and principles that are not only quantifiable but also empirically testable. This approach focuses on the physical, observable parts of reality, encouraging a deliberate and evidence-based examination of the world and its structure. It drives reasonable arguments by systematically examining even metaphysical and supernatural ideas and aims for evidence-based perspectives.

One major contrast between physicalism and materialism is that physicalism acknowledges there may be features or laws in the cosmos that we have not yet discovered or understood, or that may be more abstract or energetic than anticipated, and that these may play a role in explaining complex phenomena such as consciousness. In this regard, physicalism is more open-ended and adaptable than materialism. Physicalism accepts the possibility that the relationship between mental and physical states is more complex than known

103

reductions, but it doesn't deviate from maintaining a stance of physicality by retaining physical principles.

Physicalism, in its devotion to a naturalistic framework, is open to the possibility of even the most strange or otherworldly realities, postulating them as potential physical or chemical manifestations. This viewpoint lays the groundwork for a more in-depth investigation of "occult physics" or agnostic science - the study of hidden or undiscovered laws that may underpin these perplexing areas. Instead of thinking what exists beyond the current scope of physics as non-physical, physicalists guess that whatever exists beyond is some type of "spooky physics."

One reason being is because reports of supernatural phenomena tend to maintain physical aspects such as form, movement, and force. Some religious and supernatural doctrines align with physicalism by positing supernatural realms as advanced or concealed physical natures regulated by an unknown, or occult, physics. There is no clear distinction between supernatural and physical realms except by knowledge; the supernatural is as an outer or obscure layer of a physical existence that behaves according to principles exceeding our comprehension. This doesn't mean they're impossible to know, just that they may require different tools of analysis.

When revealed to us, these features are commonly measurable with factors that mirror design and functionality. Physicalism as a viewpoint bridges the gap between the paranormal and the physical, implying that even the most mysterious and transcendent parts of existence may ultimately have a physical basis in reality that the human brain has yet to discover. Throughout history, humanity has attempted to define and comprehend what it means to be "physical." This pursuit results in physicalism and the ideas in this book that anything that moves, expresses, implements, or has essence in any way can be considered physical.

Criticisms of Physicalism

While physicalism is a fascinating and prominent philosophical viewpoint, it is not without opponents. Several thought-provoking objections have evolved over the years, criticizing various facets of this philosophy. These critiques go into significant issues about consciousness, experience, and the boundaries of reductionism. The "knowledge" argument, the "explanatory gap," and the non-reductive perspective are among the noteworthy critiques. These critiques push us to reconsider physicalism's assumptions and boundaries, prompting a more in-depth investigation of the complexity surrounding the mind-body dilemma and the nature of subjective experience.

The Knowledge Argument

The knowledge argument, effectively developed by philosopher Frank Jackson, asserts that certain components of conscious experience, particularly "qualia" or subjective qualities of perception, cannot be fully comprehended or known through physical descriptions alone. According to Jackson's argument, even though we have perfect physical understanding of a thing, we may still lack subjective awareness of that phenomenon.

Pretend that:

1. Mary is a smart scientist who has never seen color because she has lived her entire life in a black and white environment.

2. Mary is an expert in the physiological mechanisms underlying color perception.

3. Mary escapes the monochrome space and is exposed to color for the first time.

4. When Mary sees color, she discovers something new that cannot be fully explained in terms of physical laws.

Consequently, Jackson proposes there is something about conscious experience (qualia) that cannot be entirely explained in terms of physical mechanisms.

The knowledge argument encourages us to picture Mary, a scientist, as someone who has never seen color and has only ever known shades of black and white. Despite this, Mary is a brilliant scientist who fully comprehends the physical principles underlying color perception and is well-versed in the neural mechanisms that underpin color vision. Mary, though, is eventually let out of the black and white space and gets to see color for the first time. Even though Mary previously understood everything there is to know about color perception from a physicalist standpoint, the question is whether she learns something new or obtains new information when she encounters color for the first time subjectively.

→ **Rebuttal to the Knowledge Argument**

The knowledge argument can be contested on the grounds that color vision is a unique neurochemical process with distinct information. By not experiencing the objective chemistry of color vision, Mary cannot possibly have obtained all information about color. She has technical knowledge but not the knowledge of chemical experience. One difficult aspect about Mary's situation is it's difficult to distinguish technical knowledge from experienced knowledge until such a thing like color vision is personally lived.

Color is a distinct type of knowledge based on photoreceptive experience. Color vision is the physical process anchored in the operations of the visual system. When light enters the eye, it interacts with photoreceptors, which are specialized cells in the retina. Cone photoreceptors oversee detection of different wavelengths of light and transfer that information to the brain. Every cone type is sensitive to a different wavelength range that correlates to different hues, and this is why people can be color blind of different colors at different degrees.

In ophthalmology, every part of the field is characterized by methodology and reliant on a measurability of interactions. The subjective experience of red, for example, is featured by sensations, contrast, and real chemicals. There are color vision tests and comparative analyses to measure Mary's ability to organically produce a red hue subjectively. Subjective color experiences, emotions, sensations, and perceptions are evidentially a result of physical reactions within the human body and brain.

108

The Zombie Argument

The zombie argument, developed by philosopher David Chalmers, challenges the concept that physicalism can adequately explain consciousness. He presents a hypothetical concept of a "philosophical zombie" (p-zombie) that is a sentient human lacking subjective experience. Chalmers contends that the p-zombie has an identical physical composition as a human with subjectivity, and if such a universe is logically viable, then it means that consciousness is explained by something other than the physical realm.

→ Rebuttal to the Zombie Argument

Physicalists argue, in response to Chalmers' zombie argument, that a precisely identical physical duplicate of a human would not lack consciousness since consciousness is inherently founded in physical processes. According to this viewpoint, all aspects of consciousness are generated by the brain's complicated neuronal activity, synaptic connections, and electrochemical communication. Consciousness would nonetheless appear as an inherent characteristic of the physical components in a physically indistinguishable replica. As awareness is regarded an intrinsic element of the physical universe, there would be no "consciousness gap" in such a being. However, the question over whether physicalism can fully explain consciousness or whether certain parts remain beyond simply physical explanations remains a topic of philosophical investigation.

Non-Reductionism

In relation to physicalism, non-reductionism is the philosophical position that mental states and properties cannot be entirely reduced to, or explained solely by, physical states and properties. In other words, non-reductionism contends that there are components of consciousness, subjective experience, and mental events that are not reducible to the underlying physical processes in the brain or any merely physical explanation.

The existence of qualia is one of the most important distinguishing qualities underlined by non-reductionists. The subjective, qualitative properties of conscious experience — the "what it's like" to encounter something — are qualia. Sensations such as the redness of a ripe apple, the sweetness of sugar, and the discomfort of a headache, non-reductivists propose these physical descriptions alone cannot adequately captured or explained by physics.

→ Rebuttal to Non-Reductionism

The science of subjectivity and qualia is a huge and evolving field. Subjectivity refers to a personal and distinct essence of conscious experience, which may differ from person to person. Qualia encompasses the subjective and qualitative traits of experiences like food flavors and painful sensations. While these occurrences pose philosophical problems, the physicalist viewpoint argues that they can be successfully comprehended via empirical science.

Importantly, physics is an ever-evolving field of knowledge not a closed system. Physicalists anticipate that any findings about consciousness will eventually be revealed to be physical in nature. This viewpoint is based on the premise that the fundamental rules and principles that control the cosmos apply cross-dimensionally, so even the most supernatural aspects of existence and consciousness are in the realm of physics.

Subjectivity, by self-report, is physical. It's an account of shifting sensations (motion), transforming thoughts (state change), expressive symbols & sceneries (form and motion) — precisely the topics that physics covers. While we may not always understand the intricate workings of cognition and subjectivity, evidence points in the direction of abstract conversations like God and consciousness being completely in the study realm of physical science.

The Explanatory Gap "Hempel's Dilemma"

Hempel's Dilemma raised by Carl Hempel is a big challenge to physicalism that focuses on two key points: "explanatory gaps" and of "multiple realizability." Explanatory gaps imply that physicalism struggles to completely explain certain mental events in terms of purely physical processes, such as consciousness and qualia. Multiple realizability indicates that many physical systems can produce the same mental state. Hempel's Dilemma urgently questions whether physicalism can appropriately address the intricacies of mental experiences.

→ Rebuttal to Hempel's Dilemma

A physicalist defense against Hempel's Dilemma is that, while the explanatory gap is accurate, it is not definitive evidence against physicalism. Instead, it serves as a sober reminder that physicists' existing understanding is inherently limited. The concept of the "non-physical" is never clearly defined by Hempel or expressed in a way that gives any meaningful alternative to physics or the physicalist paradigm.

According to physicalism, everything, including mental occurrences, is ultimately anchored in the physical universe. As a result, the challenge offered by the explanatory gap emphasizes the need of continually reflecting on the urgency of developing a more comprehensive physics — one capable of accounting for the complexity of consciousness, qualia, and subjective experiences. It is a call to review and correct any gaps in our present understanding of physical processes and their relationship to mental states. In essence, while the explanatory gap raises serious concerns, it mostly highlights the importance of continuing the work to explore and enhance physics rather than rejecting it.

Reasons People Might Reject Physicalism

Critics of physicalism often argue that not all aspects of human experience can be reduced to the physical world. One of the most well-known alternatives is dualism, the belief that the mind and body are separate entities. Dualists contend that mental states—such as thoughts, feelings, and consciousness—cannot be fully explained by physical processes in the brain, raising the issue of how mind and body interact.

Another challenge is the concept of qualia, which refers to the subjective, individual experience of sensations. For example, the way we experience the color red or the taste of salt is unique to each individual and seems to defy physical explanation. This leads to the argument that physicalism cannot account for the richness of consciousness.

The issue of free will also poses a problem for physicalism. If all human behavior is determined by physical causes — such as brain chemistry and biology — some argue that this leaves no room for free will or personal agency. Without free will, our actions would be mere reactions to physical stimuli, undermining our sense of moral responsibility.

The concept of emergence further complicates the picture. Some argue that phenomena like consciousness arise from complex systems but cannot be reduced to the physical components of those systems. For example, while individual neurons are simply cells firing, the subjective experience of consciousness is not something that can be fully explained by looking at individual parts in isolation.

On the other end of the spectrum is idealism, which posits that reality is fundamentally mental, and that the physical world is an illusion or a construct of the mind. Idealists argue that consciousness is the primary substance of the universe, which is in direct contrast to the physicalist view that only the material world is real.

Another concern is personal identity. Many reject physicalism because they see the self as something non-physical, a soul or essence that persists over time, which cannot be explained through physical processes alone. This raises the problem of mental causation: how can non-physical mental states cause physical events, like the movement of muscles or the firing of neurons?

Metaphysics

Metaphysics is a branch of philosophy that, like physicalism, explores the nature of reality and existence. However, while physicalism is concerned with the material world and empirical science, metaphysics addresses questions that lie outside the scope of physics, empirical observation, and scientific study.

Though it may use terminology similar to that of the sciences, metaphysics is not a science. It cannot replace physical science because it lacks the rigorous scientific method. Instead, metaphysics delves into abstract aspects of existence that science cannot yet explain — either due to limitations in our tools or our own human capacity. For instance, some phenomena are beyond our detection because of the limitations of our sensory perception, the scope of our scientific instruments, or the inherent diversity in human biology and environmental constraints.

This is where metaphysics plays a crucial role: it serves as a philosophical space for exploring questions and ideas that are too obscure or intangible to be investigated through scientific means. Though metaphysics lacks the empirical testing found in science, its purpose is not to challenge or replace science but to complement it by offering a space for contemplation about the nature of reality that may not yet be accessible to scientific inquiry.

While metaphysics is abstract and often speculative, it is still valuable for posing questions about the unknown. As certain metaphysical topics become more empirically observable or relevant to our senses, they may transition into the realm of scientific study. The rigor of scientific methodology applies when a topic can be tested, observed, and falsified.

It's important to understand that the rules governing philosophy differ from those of science. Philosophy allows for more abstract, open-ended exploration and questioning, whereas science is grounded in empirical data and testing. Metaphysics explores the boundaries of physics, asking questions about phenomena that may exist beyond our current understanding. This space remains open to curiosity and is intended to encourage reflection on what might lie outside of what we can currently observe or test.

While metaphysics doesn't rely on practical observation for verification, this doesn't mean that metaphysical phenomena are non-physical. Instead, it suggests that we simply don't have the tools or the knowledge to fully comprehend them yet. In this way, metaphysics aligns with agnosticism — it operates in areas of uncertainty, where definitive answers are elusive, and it refrains from making strong claims about the nature of reality.

Exercise #12: Discerning Metaphysics from Physics

To begin, physics is the natural science concerned with understanding the fundamental principles that govern the physical universe. It focuses on measurable phenomena such as energy, matter, and the forces at play in the universe, and relies heavily on observation, experimentation, and mathematical modeling. Physics is rooted in empirical methods that aim to explain what exists, how it works, and why it behaves the way it does.

In contrast, metaphysics is a branch of philosophy that tackles questions beyond the scope of physical science. It addresses abstract concepts like existence, reality, causality, and time, investigating why there is something rather than nothing, and what it means for something to exist. Metaphysics often explores concepts that cannot be directly observed or measured, making it distinct from physics, which deals with tangible, empirical data.

Physicalism is a philosophical viewpoint that aligns closely with the scientific method. It posits that everything that exists is either made of physical matter or can be fully explained by physical laws and interactions. For physicalists, all phenomena — whether mental, emotional, or conscious — can be reduced to physical processes, like brain activity or molecular interactions. In this worldview, consciousness, emotions, and even thoughts are byproducts of physical systems.

This perspective mirrors the approach of physics, which strives to understand and explain the universe using observable, testable, and quantifiable data. Physicalism is aligned with the scientific method in that it assumes all phenomena can ultimately be described by physical laws. This concept is especially relevant in fields like neuroscience, which seeks to understand mental processes through brain activity and chemical reactions.

At the core of physics is the scientific method, a systematic process for discovering knowledge through observation, experimentation, and the formation of hypotheses that can be tested and refined. In physics, facts are gathered through sensory experiences and experiments that produce measurable data. These data help scientists develop models and theories that explain how the physical world operates.

The scientific method stands in contrast to metaphysical inquiry, which does not rely on empirical testing but rather philosophical reasoning and conceptual exploration. While physics focuses on what is measurable and observable, metaphysics delves into what is conceivable and speculative — such as the nature of being or the existence of parallel realities.

To understand the distinction between physics and metaphysics, let's consider the nature of questions asked in each field. In physics, scientists might ask, "What are the physical properties of light?" or "How does gravity affect objects in motion?" These questions are grounded in observable facts and can be tested using controlled experiments. Physics, therefore, focuses on the mechanisms of the physical world.

Metaphysical questions, however, are more abstract. For example, "What is the nature of time?" or "Do we have free will, or is everything predetermined?" cannot be directly observed or tested. Instead, metaphysical inquiries involve pondering ideas that cannot be empirically validated. In this way, metaphysics frequently deals with questions about existence and reality that remain outside the scope of scientific testing and are often subject to philosophical debate.

Both physics and metaphysics serve as essential ways of understanding the world, though they do so in different manners. Physics explains how things work based on observable evidence, while metaphysics asks why they exist in the first place and what the deeper nature of reality might be. Despite their differences, the two are not mutually exclusive. While physics focuses on the tangible, metaphysics provides room for exploring the intangible aspects of human experience—such as the mind, consciousness, and existence itself.

The idea of physicalism ties these two domains together. By proposing that everything, even consciousness and abstract ideas, has a physical basis, physicalism attempts to reconcile the philosophical inquiries of metaphysics with the empirical methods of physics. This interplay between abstract philosophical speculation and rigorous scientific inquiry invites us to consider the full spectrum of reality, from the physical to the metaphysical.

To solidify your understanding, take a moment to reflect on the following questions:

Which aspects of reality do you feel can be explained purely by science, and which might require metaphysical contemplation?

How does physicalism challenge or support your view of the universe and existence?

Can you think of any real-world phenomena that blur the line between physics and metaphysics?

SECTION II:

ENCHANTED

MECHANICS

CHAPTER 4

THE MAGICKIAN

The last vestiges of fall held fast in the Appalachians, with the skeletal trees reaching into the overcast sky, while the last of the goldenrod stood defiant against the encroaching winter. Annie, wrapped in her scarf, cast her gaze over the ridge where the solar panels lay, their surfaces dulled by a light frost, yet gleaming with potential. This project, though not yet fully operational, symbolized a new era for the region, one where the mountains could harness their own energy to combat the unpredictable weather patterns that had become all too frequent.

Lucian, with his weathered face and hat askew, observed the scene with a mix of curiosity and skepticism. "I understand the sun," he remarked, his voice carrying a hint of doubt, "but all this talk of magick and technology? How's it supposed to do anything about the floods?"

Annie's laughter was gentle. "It's not magic in the traditional sense, Lucian. It's about capturing and utilizing energy, much like how nature works. Solar panels are just one part of a larger ecosystem of change."

Lucian pondered her words, his skepticism giving way to a cautious interest. "And this magick everyone mentions? Is it part of this... ecosystem?"

"Absolutely," Annie affirmed. "Magick, in this context, is about recognizing the inherent potential in everything. It's the energy, the transformation that happens when elements combine in new ways. The magickians? They're just scientists with a poetic view of the world, using science to weave solutions."

"Well," Lucian said, nudging a stone with his boot, "sounds like I need to meet these magickians of yours."

The Gathering

As twilight settled, Lucian followed Annie into the heart of the woods, where the pine trees formed a natural sanctuary. They emerged into a clearing alive with the glow of campfires and the buzz of modern machinery. The crowd was eclectic, their appearance a blend of the arcane and the avant-garde, with neon hair clashing against the practicality of their work gear.

Lucian's initial impression was one of bemusement. "Looks like some kind of underground club," he whispered.

Annie grinned. "They're just a bunch of innovators. Don't mention anything about government support, or you'll get an earful."

Stepping into the light, Lucian was introduced to a world where technology and nature intertwined. A drone danced overhead, its lights painting ephemeral patterns on the ground. A chemputer, an alchemy of sorts, was in operation, with a woman with green hair overseeing its functions. Others were modifying a vehicle designed to navigate both land and water, its parts muddy from recent use.

"Lucian," Annie introduced, "these are the magickians."

A man with glasses and a jacket adorned with tech insignias approached. "I'm Kael," he said, extending a hand. "You must be Lucian. We're glad you're here."

Lucian shook Kael hand, still processing the scene. "Kael, haha! This is... different from what I expected."

Kael chuckled. "Kael with an "A-E" not sea kale. Magickians aren't all hippies. We're often misunderstood. I'm just trying to leverage the universe's own patterns to improve life here."

As night progressed, Lucian listened to discussions ranging from using drones for community vigilance and aid during disasters to the transformative power of assembly theory in medicine. Caleb explained how their chemputer could potentially transform local resources into necessary drugs or materials, envisioning a future where each Appalachian community could be more self-reliant.

"How do you convince the locals to embrace this?" Lucian asked, his initial skepticism now laced with interest.

"That's where people like you come in," Annie said. "You know the community's pulse. You can help us show that this isn't just change for change's sake, but a way to preserve what we love about our home while adapting to new realities."

Lucian watched the drone, its lights flickering like stars. He thought of the resilience of the people here, their connection to the land,

and began to see how tradition and innovation could coexist, like moths and flames.

"Alright," he conceded with a grin, "let's see what this magick can really do for Appalachia."

The Spirit of Transformation

The evening unfolded with demonstrations of technology that seemed almost mystical in its application. Lucian tried the self-driving amphibious vehicle, the experience both thrilling and enlightening. He watched as the chemputer synthesized compounds, a process so precise it bordered on the miraculous.

In this blend of the old and new, surrounded by the hum of machinery and the warmth of community, Lucian felt a shift within himself. "Magick," Annie whispered, "is all around us, Lucian. You just have to open your eyes to see it."

And as Lucian looked out into the night, he realized that perhaps the spirit of transformation was not just in the technology but in the willingness to adapt, to blend the wisdom of the past with the tools of the future. It was movement. Maybe, just maybe, this was the start of something truly transformative for his beloved Appalachia.

History of Magick

Magick, within human philosophy, dances across a broad spectrum of interpretations, all circling the theme of transformation. While many traditions spotlight the pivotal role of human will and intention in sculpting reality, another perspective casts magick as a timeless dance between essence and existence. This notion suggests that magick's spirit might pulse independently of human consciousness or other sentient entities, existing as a fundamental pattern in the fabric of reality.

Agnostics and atheists, free from the chains of dogma or the allure of panpsychism, can forge innovative paths in physical magick. Meanwhile, those of faith can leverage Jangled Jester's definition to bolster a physicalist spirituality. Here, magick is seen as the architecture of existence, where spirit is not just a passenger but part of the blueprint. To be a magickian, then, is to engage deeply with these structures, to study their patterns, and to manipulate their formulas with precision.

According to Jangled Jester, magick is not merely an art or a science but an intricate fusion of both. This vision underscores the dynamic interplay between the tangible evidence of work and the boundless realms of creativity. It celebrates inquiry and imagination as twin engines driving mastery in magick, promoting a symbiosis of structure and operation. Magick, in this view, is woven into the very sinews of reality, inviting practitioners to not only explore but to embody and foster transformation. It marries the methodical rigor of science with the expressive freedom of art, all while pondering the design and formula of our universe.

Whether one's journey leads towards supernatural transcendence or focuses on earthly matters like sustainability and extending human life, magick remains a tapestry rich with both scientific inquiry and artistic expression. It honors the individual's spiritual or secular journey, be it through divination or empirical prediction. The ultimate aspiration for all magickians is to adeptly wield both art and science in their quest.

On the question of whether magick harbors evil or antihuman elements, one must delve into the potential of form and function to be turned against us. Clearly, they can. If magick aligns with the principles of form and function, understanding how these can be misused becomes crucial. For those without religious affiliation, evil isn't an external force but a product of the human condition, arising from our own biochemical makeup. Thus, magick encapsulates the full spectrum of existence's possibilities, presenting itself as an echo of creation, inherently transformative yet fundamentally neutral. It's a pattern, replicated throughout existence, inviting us to engage with its potential, for good or ill, in the grand design of things.

The Magickian

Magick is not just about being entranced by the mystery but actively seeking to understand it. This pursuit of knowledge over mere wonder defines the magickian. Here, the balance between enchantment and inquiry is paramount. To be perpetually dazzled without question can dull the senses to deception, whereas an excess of doubt can alienate one from the world's wonders. The magickian thus navigates this delicate balance, fostering both an appreciation for life's magic and a critical analysis of its workings.

For the physicalist magickian, this philosophy manifests in a profound respect for the body's constant flux. They see themselves as dynamic entities interacting with both the complex, purposeful world and the indifferent, inanimate magick that surrounds us, which remains oblivious to our consciousness. This perspective, tinged with absurdity, fosters a deep link between our corporeal existence and our creative, inquisitive spirit. It's an integrative view that harmonizes the living with the non-living, the magical with the everyday, and the profound with the prosaic.

Embracing the identity of a physicalist magickian means recognizing and celebrating the transformative essence of existence. In this role, one unveils the enchantment that lies dormant in the mundane. By training oneself to see beyond the surface, to perceive the ordinary as a canvas of transformation and energy, the magickian not only enriches their own experience but also has the power to guide others towards this revelation. This approach transforms life into a quilt of moments, each holding the potential for profound insight and change, making every interaction, every observation, a thread in the magickal fabric of reality.

Exercise #13: Identifying Physical Magick

Objective: To enhance your perception of the world by breaking down objects, events, and situations into their basic components of form and function, thereby revealing the magick inherent in the physical.

Instructions:

Choose something mundane from your environment. It could be a coffee mug, the act of walking, a tree outside your window, or the process of cooking dinner.

Begin by observing the form of your chosen subject. Form refers to the physical attributes:

What does it look like? Describe its silhouette, its contours.

What is it made of? Consider the properties of the material — its texture, color, flexibility, or rigidity.

How are its parts organized or connected? Look at the arrangement, the design, or the internal structure.

Now, delve into the function of the subject:

What is its primary use or role? If it's an event or action, what does it achieve?

How does it work? Describe the process, the motion, or the interaction.

What change does it bring about? Consider the immediate or long-term consequences of its function.

Reflect on how form and function intertwine:

How does the object or situation transform energy, matter, or even emotions?

How does its form support or enhance its function? Or perhaps, how might it limit or challenge its function?

What other functions could its form support? Could it be repurposed or imagined in a new way?

Now, use this newfound understanding to inspire action or thought:

If you could redesign this object or situation, how might you do it to enhance its function or introduce a new one?

Create something inspired by this analysis. It could be a drawing, a poem, or even a new use for the object.

Identify a problem related to this object or situation and use your understanding of its form and function to propose a solution.

Write down your observations and insights. How does seeing the world through this lens of form and function change your interaction with it?

Consider sharing your findings with someone. Discuss how breaking things down in this way can lead to innovation or a deeper appreciation for the ordinary.

Magick Example:

Object: A simple pencil.

Form: Cylindrical wood, graphite core, eraser on top, painted in yellow, hexagonal sides to prevent rolling.

Function: Writing or drawing, erasing mistakes, providing grip.

Magick: The pencil transforms thoughts into visible marks, embodies human creativity through its simple mechanics. Its form aids in easy handling, while its function allows for correction and creativity.

Deviation from Aleister Crowley's Concept of Magick

Aleister Crowley's well-known definition of magick from the late 19th and early 20th century frames it as "the Science and Art of causing Change to occur in conformity with Will." This perspective tightly binds magick to intention, consciousness, and an arguably panpsychist or idealistic view of nature, where all entities possess some form of mental quality. However, if we conceive magick as "the design of function," we shift the focus away from mental attributes to embrace the inherent properties of structure and process, inclusive of those that might objectively lack consciousness or intent. While panpsychism offers one lens, physicalists can also engage with magick by exploring the neutral, structural designs and processes of nature.

Jangled Jester's notion of magick transcends the need for intention, suggesting that magick can be a property embedded within inanimate designs and the mindless operations of entities like the sun or water. Here, art becomes a manifestation of design principles, and science, evidence of work. Even without divine inspiration, the sun and water have intrinsic structures and functions that give them a unique 'spirit'. If parts of existence evolved without a sentient overseer, then their study through form and function is not only valid but essential. Magick, in this view, is the elusive, transformative 'energy' it's reputed to be because it embodies the diverse capabilities of forms to change through function. Although intention represents a facet of magick, it's not the sole essence;

there are also unintentional, non-cognizant aspects of existence that merit inclusion in magickal exploration.

Jangled Jester's approach acknowledges intention as one variety of magick but does not elevate willpower as its pinnacle. Intention can be explored both scientifically, through cognitive processes like goal-setting, and metaphysically, where interpretations might delve into less tangible realms. Scientific inquiry into intention can involve cognitive psychology or neuroscience, examining how the brain translates intention into action. Metaphysical views might introduce non-physical or supernatural elements, which might not resonate with agnostic atheists due to their lack of empirical grounding. Both religious and non-religious metaphysical ideas are worth considering, but they should be approached with a commitment to evidence, logic, and reason.

Some might dismiss magick as meaningless, but contemplating the significance of form and function, even in relation to the supernatural, can be enlightening. Viewing supernatural phenomena within a physical framework allows for understanding these experiences in terms of known physical laws. For instance, transcendence can be seen as a physical elevation or transformation in underlying forces. This approach helps in defining the shapes and properties of even the most abstract entities like ghosts or gods. Physicalism can complement religious ideas, appealing to a wide range of spiritual and non-spiritual practitioners, fostering interdisciplinary collaboration in magickal practices.

Ultimately, how one chooses to engage with magick is a personal decision. Not all aspects of magick are beneficial, relevant, or honest. While some magickians might see magick as the deliberate act of a sentient creator, others might view it as an intrinsic relationship between existing structures and their processes, devoid of religious context. Whether religious or nonreligious, physicalist magick seeks to measure and understand experiences, including those traditionally considered supernatural.

Crowley's definition narrows magick to the domain of willfulness, suggesting either an elite practice or a panpsychist perspective. In contrast, Jangled Jester's physical definition of magick does not presuppose willful structures or operations. Magick exists irrespective of our interpretations. While will can result from magick, it isn't the origin of magick. Physicalist magick focuses on the broader design and operational principles that shape the essence and evolution of existence, rather than fixating solely on will.

Design in Magick

Design holds a pivotal role in magick, not only in the deliberate creations of human endeavor like graphic design but also in the inherent patterns found in nature that might not have a conscious architect. While human design is intentional, there exist natural designs, such as the veins on a leaf or the smooth contours of river stones, that evolve without deliberate intent. The significance of design in magick lies in its suggestion that all forms in existence inherently tend towards structure and placement. By recognizing the physical characteristics associated with design, magickians can harness their awareness to develop tools and methodologies for understanding and interpreting the complex patterns we see both in the natural world and within our own being. Engaging with the design principles that govern both intentional and spontaneous patterns allows magickians to navigate and utilize the transformative potential embedded within them.

Design Principles in Magick

Alignment: Strategically placing elements to suggest order and structure, which can aid in creating a focused magickal intent.

Hierarchy: Ranking elements by importance to direct the flow of energy, guiding the viewer's or practitioner's focus to where it's most needed.

Contrast: Using stark differences to highlight transformations or to symbolize duality and change within magickal workings.

Repitition: Employing repetition to reinforce intent, create stability, or invoke the power of rhythm and ritual.

Proximity: Positioning related elements near each other to strengthen associations or to symbolize interconnectedness in magickal constructs.

Balance: Distributing visual or symbolic weight to achieve harmony, which can be crucial in maintaining equilibrium in magickal energies.

Color: Leveraging color theory to influence the emotional and spiritual resonance of a magickal operation, aligning with the symbolic meanings of colors.

Space: Manipulating space to enhance the perception of depth or to emphasize isolation, which can be used to focus or expand magickal intent.

Emphasis: Highlighting key components or symbols to ensure their prominence in the magickal act, akin to focusing will or energy.

Proportion: Scaling elements to reflect their significance or to embody concepts like growth, reduction, or balance in magickal symbolism.

Rhythm: Creating a flow that can mimic the cycles of nature, the rhythm of breath, or the heartbeat, useful in timing and pacing magickal work.

Pattern: Utilizing patterns to represent cycles, continuity, or the fractal nature of existence, often used in magick to connect with universal energies.

Movement: Guiding visual or energetic flow to direct the practitioner's focus or to symbolize progression or transformation.

Variety: Introducing diverse elements to keep the practitioner engaged, preventing stagnation in magickal practice, and encouraging adaptability.

Unity: Ensuring all components work in harmony to convey a singular, powerful magickal statement, enhancing the potency of the spell or ritual.

Lines and Shapes: Employed to define boundaries, contain energies, or to direct the flow of power, with shapes often holding specific symbolic meanings in magick.

Exercise #14: Identifying Design Principles

Select a location within your community where you can easily observe various elements of design. This could be a park, street, market, or any public space that interests you.

Take a leisurely walk through the chosen location, paying close attention to your surroundings. Observe the objects, structures, and natural elements present.

As you walk, identify and note down instances where you recognize the 17 design principles we discussed earlier. Look for patterns, shapes, alignments, contrasts, repetitions, and other principles that are evident in your environment.

Document your observations using a notebook or smartphone. Take photographs or make sketches to illustrate the design principles you've identified.

After your observation walk, sit down and reflect on the design principles you've found. Consider how these principles contribute to the functionality, aesthetics, and organization of your community.

Function in Magick

Function is crucial in magick as it serves as the conceptual framework that explores the intricate dynamics governing the transformation and interaction of forms. While design focuses on the structural components of forms, function delves into how these components interact, evolve, and manifest their essence or 'spirit'. In a mathematical sense, functions depict the relationship between variables, much like how functional interactions in nature reflect the essence of physical structures.

This concept is exemplified in the human body, where myriad functions underpin its complexity, from biochemical processes to physical movements. The essence of functionalism lies in its diversity and dynamism, attributing to all entities, animate or inanimate, a vibrant quality of spirit. This spirit can be understood as the capacity for complexity and change. For instance, the spirit of a flower arises from the seamless integration of its parts — stem, roots, petals — into the holistic entity we recognize as a 'flower.' Similarly, the human spirit emerges from the intricate interplay of bones, muscles, hormones, and more within the body's sophisticated architecture.

Classes of Functions

Fractals: These are patterns that exhibit self-similarity at different scales, observed in nature like the branching of trees or the structure of snowflakes. Fractals represent efficiency in space-filling and resource distribution.

Symmetry: Nature often showcases symmetry, from the radial patterns in starfish to bilateral symmetry in humans, which facilitates balance and functional coherence.

Fibonacci Sequence: This sequence, where each number is the sum of the two preceding ones, manifests in natural growth patterns, such as the spiral arrangements in plants, representing an optimized growth strategy.

Optimal Packing: Seen in structures like honeycombs or the atomic arrangement in crystals, optimal packing maximizes space usage and efficiency in nature.

Self-Organization: Ecosystems and other complex systems self-organize to achieve stability, as seen in ecological balances, nutrient cycles, and population controls.

Natural Selection: This biological process, where traits beneficial for survival and reproduction are favored, drives functional evolution and adaptation in species.

Mimicry and Camouflage: These adaptive strategies enhance species survival by blending into environments or mimicking others, illustrating function through deception or concealment.

Symbiosis: Interactions where species benefit from each other, like the relationship between bees and flowers for pollination, or between plants and mycorrhizal fungi, showcasing interdependence as a function.

Ecosystem Succession: The progressive change in species composition over time, leading to stable ecosystems, is a function that illustrates development and adaptation in the environment.

Chemical Bonding: The way atoms bond to form molecules dictates their function, crucial for life at the molecular level.

Feedback Loops: These systems regulate conditions through loops where changes in one direction lead to counteractions, maintaining balance, like temperature regulation in organisms.

Predator-Prey Dynamics: The mathematical modeling of population changes in predator-prey relationships, which can maintain ecological balance or lead to cycles of population growth and decline.

Trophic Levels: The structured feeding relationships within an ecosystem, with energy transfer from one level to another, underpinning ecological function.

Biogeochemical Cycles: The cycling of elements like carbon, nitrogen, and water through the biosphere, lithosphere, and atmosphere, essential for sustaining life and environmental stability.

Hydrodynamics: The study of fluid motion, which governs phenomena from ocean currents to the flow of blood, illustrating function through fluid dynamics.

Geological Processes: These include processes like plate tectonics, which not only shape the Earth's surface but also influence climate, habitats, and ultimately, the distribution and evolution of life forms.

Exercise #15: Identifying Function in Daily Life

Start a functional analysis notebook in which you document numerous elements of your day, concentrating on activities, objects, or processes. Take note of their structural foundations and how they function or interact.

Select a single day or period to observe closely. Determine the relationships between diverse elements by taking into account their dynamic states, changes, and reactions to various stimuli or influences.

Consider how the correlations you identified can be stated quantitatively, similar to the concept of functions in mathematics.

Extend the exercise to your own body. Consider the different chemical, physical, and biological functions that contribute to its complexity. Consider how these functions combine to keep your body in good health.

Draw attention to examples of functional diversity in your environment. Identify things, systems, or processes that exhibit a variety of functions, reflecting the diversity present in nature.

Consider the essence or spirit of function in the elements you've observed. Consider how their unique and dynamic characteristics contribute to a sense of spirit or life in your daily interactions.

Take into account the comprehensive integration of functions. Investigate how diverse elements work in harmony to form a larger

146

whole, whether in the operation of a device, the operation of a system, or the functioning of your own body.

More History of Magick

Magick, with a "k," has its roots in the scholarly works of Heinrich Cornelius Agrippa, a 16th-century polymath renowned in occult circles. The distinctive spelling might have resonated with Aleister Crowley, who later adopted it to differentiate his esoteric practices from common stage magic. However, it's essential to note that Agrippa documented the use of 'k' in translations well before Crowley was born in the 19th century. This intentional spelling variation marked a significant shift in the conceptualization of magick.

Traditionally, magick is regarded as a spiritual discipline aimed at forging connections with transcendental, supernatural realms, mystic forces, ancestral knowledge, and divine objectives. Practitioners seek to harness magick as a means to manifest their deeper aspirations, embarking on paths of self-discovery that lead to enlightenment and wisdom. While some historical and contemporary uses of the term omit the 'k', the inclusion of this letter often serves to differentiate the esoteric, transformative practices of magick from entertainment-based magic tricks and illusions.

In a poetic sense, both magic and magick act as gateways to a profound understanding of existence, allowing individuals to view the world with a sense of wonder and reverence. These practices transcend the ordinary, blending the everyday with the mystical, and encourage an exploration of our interconnectedness with the cosmos. Magick,

148

specifically, positions itself as a philosophical stance that inspires active engagement with the fabric of life, urging individuals not to remain passive observers but to recognize their role as active participants in the continuous dance of reality. Here, everyone is seen as a natural-born magickian, endowed with the capacity to shape and influence their reality through intention, thought, and action.

To be a natural-born magickian means to possess an inherent link to the cycles of existence and the power to mold one's environment through conscious engagement. Such individuals are tuned to the enchanting elements woven into the fabric of daily life, inviting us to embrace our roles as shapers of our destinies, with the tools to perceive, interact with, and alter the tapestry of existence. This perspective celebrates the extraordinary within the ordinary, encouraging individuals to explore life's mysteries and wonders with self-initiated purpose and curiosity.

Historically, the practice of magick has been deeply entwined with recognizing humanity's intrinsic bond with nature. From indigenous peoples to the ancient civilizations of Egypt and Greece, and through medieval and Renaissance esoteric systems, humans have been seen as active agents in the cosmic dance of life. Through rituals, divination, and symbolic acts, these cultures cultivated a profound awareness of the symbiotic relationship between humans and the environment, acknowledging the mutual transformative power at play.

149

In medieval and Renaissance Europe, the Western esoteric traditions not only sought to solve the philosophical enigmas of life but also adopted a practical, hands-on approach to magickal arts. Alchemists, hermetic philosophers, and mystics explored nature's secrets through a fusion of art, craft, process, chemistry, and early scientific methods. Grimoires became the repositories of this knowledge, outlining magical practices in systematic steps to be handed down through the ages. This comprehensive approach to magick underscores humanity's persistent endeavor to tap into its latent capabilities, merging theoretical insights with practical application in a quest to interact with the mystical forces of the universe.

Exercise #16: Get to Know a Historic Magickian

Choose a historic magickian whose life and contributions pique your interest. It might be Hermes Trismegistus, John Dee, Aleister Crowley, or any other influential historical magickian.

Conduct extensive research on your chosen magickian. Investigate their biography, writings, and teachings, as well as the historical circumstances in which they lived. Document important parts of their lives, such as their influences, notable works, and the cultural or philosophical context of their time.

Investigate the magickian's procedures firsthand. If they left behind specific rituals, spells, or procedures, try your best to reproduce them. This hands-on experience connects you more viscerally to their approaches and insights.

Think about your discoveries and experiences. Consider how the life and teachings of the magickian resonate with your own beliefs and practices. Determine which elements are particularly appealing or relevant to your spiritual path. Incorporate these revelations into your personal understanding of magick and spirituality.

The Future of Magick

Looking ahead, the future of magick presents an expansive canvas for individuals to delve into the physicalist depths of our existence and to revel in the splendor of the material world. This era signifies a transition from viewing complex scientific disciplines like organic chemistry as dry or daunting. Magickians are now poised to reimagine organic chemistry as a sophisticated and potent magickal language — one that allows for the deciphering and crafting of matter's various expressions. This transformation not only elevates the mundane to the realm of the marvelous but also facilitates a vibrant integration of scientific inquiry with magickal exploration.

As we venture into a new epoch, physicalist magick ignites the potential for extraordinary developments that transcend our current planetary confines. The application of magick in facilitating human colonization of the Moon and Mars heralds an ambitious journey — one that extends into the vastness of space and its myriad enigmas. This exploration holds the promise of unveiling innovative solutions to longstanding challenges such as eradicating most diseases, extending human lifespans, and creating sustainable ecosystems where new life forms can flourish in health and happiness. During this audacious journey, the symbiosis of magick and science emerges as a guiding light, leading humanity toward breathtaking revelations. In this new era, the boundaries between what is known and the mysteries yet to be uncovered blur, heralding a future brimming with potential and wonder.

TSOHPM

Exercise #17: Envisioning the Future of Magick

What role do you see for magick in humanity's exploration of the Moon, Mars, and beyond? Investigate how magick might aid in the construction of sustainable habitats and the finding of novel solutions to difficulties in these cosmic frontiers.

Imagine the harmonic mingling of scientific inquiry and magickal research. How may these two seemingly disparate realms work together to open up new possibilities, moods, and transform our view of reality, as well as propel mankind toward a future where the lines between science and magick cross?

How do you see magick influencing personal change and progress in the future? Consider how magick could be a catalyst for people to live more fulfilling lives, founded in a profound connection with the physical world and actively partaking in the mysteries of life.

Brainstorm discoveries that may come at the junction of magick and scientific advancements. Consider the advancements in healthcare, longevity, and sustainable living that magick could help to stimulate in partnership with scientific pursuits.

CHAPTER 5

THE MAGICIAN

The air in Frostbite Holler was sharp, laced with the scent of burning tobacco from old-fashioned smokers and pine smoke, all underlaid by the crispness of new snow. Annie stood at the entrance of this transformed tobacco farm, now a pulsating winter carnival. The rolling hills of Appalachia, cloaked in white, vibrated with the raw energy of both locals and visitors under a star-pierced sky.

This wasn't your standard holiday festivity. Lucian's vision had morphed this into a raw, edgy blend of Appalachian tradition and futuristic innovation, with a hefty dose of punk rock attitude. The festival grounds were alive with flickering light shows from drones that mimicked the northern lights, while heavy metal music from multiple stages filled the air with its defiant beats. Robotic servers, styled after figures from local legends, moved through the crowd, serving not just cider, but also moonshine in mason jars, adding to the rebellious atmosphere.

For Lucian, Frostbite Holler was a declaration of intent. Appalachia, often seen as the forgotten backwoods, was here to claim its place in the modern world. This event was set to ignite a technological renaissance, underscored by the music's raw energy.

The heart of the festival was the main stage, a structure of ice and metal, where the first act, a techno-illusionist named Zalene, took the stage. Her performance was a dazzling mix of traditional sleight-of-hand with augmented reality, pulling not just stars but entire galaxies from the frosty air, perfectly timed with the pounding bass.

Annie wandered through the festival, where technology was showcased amidst the rustic charm. Here, visitors could undergo rapid health scans in mobile units, providing advanced medical care to the remote mountain communities, all set to the backdrop of punk and metal tunes. Dome habitats, simulating life on a distant icy moon, attracted crowds, where children and adults alike imagined life among the stars, the music giving everything a surreal, rebellious edge.

Yet, the spirit of Appalachia was ever-present. Workshops on survival skills were held in tents made from camouflaged fabric, seamlessly integrating modern security methods like drone-assisted surveillance, led by local experts keen on preserving old ways while embracing the new.

At the festival's core, where Lucian and Annie's vision truly unfolded, was a sensory room where visitors could dive into virtual reality landscapes or engage with brain-computer interfaces, playing mind-controlled games, the heavy metal soundtrack enhancing the experience.

The festival's highlight was a raffle for a lunar trip, symbolizing the boundless aspirations of the community. A holographic moon floated above, casting a silver glow over the crowd, pulsating with the rhythm of the music.

As the night deepened, Annie and Lucian sat by a fire, the smoke from tobacco mingling with the cold air, sharing a moment of tranquility amidst the vibrant festival.

Annie stood amidst the pulsing heart of Frostbite Holler, the festival's clamor echoing like a heavy drumbeat in her chest. She leaned in close to Lucian, her voice barely a whisper against the backdrop of electric guitars and the crowd's roar.

"Here's the thing about magic," she began, her eyes gleaming with a mix of mischief and insight. "It's all about fooling the eye, right? Making you believe in something that isn't there. But that's not what we're about, Lucian. Not really."

Lucian's attention was fully on her now, the festival's wild energy somehow amplifying the significance of her words.

"Magick, the way we're doing it here, it's not about fooling anyone," Annie continued, her gaze sweeping over the festival's chaotic beauty. "It's about pushing against what we've been told is possible. It's about looking at this mess, this beautiful, noisy chaos, and seeing not just a party, but a playground for the mind."

Lucian nodded, the concept of magick taking on a new, more profound meaning.

"It's heavy, isn't it?" Annie said, her voice rising slightly. "All this noise, all this energy. It's like trying to find a quiet moment in the eye of a storm. Magick, real magick, it doesn't just entertain; it disrupts. It's about making the mundane magical, about seeing the potential in the chaos."

158

She paused, watching as a group of kids played a game where they controlled drones with their thoughts, their laughter a sharp contrast to the metal music. "But it's in this chaos, this noise, where dreams are born. Where we dare to think about colonizing moons or whatever else seems absurd. Because magick isn't just about what we do here tonight; it's about what we aspire for tomorrow."

Lucian looked around, the festival's surreal blend of technology and tradition suddenly a canvas for transformation. "You're saying we're planting seeds here. Seeds of wild ideas, of daring to dream beyond our limits."

"Exactly," Annie's smile was audacious. "We're not just showing off tricks or gadgets. We're telling people, 'Hey, look. The world as you know it? It's just a starting point. What if we could live on an icy moon? What if we could bend reality, not with a trick, but with determination and imagination?'"

Lucian chuckled, the festival's energy now feeling like a catalyst for something bigger. "So, this isn't just a festival. It's a rebellion against the ordinary, a call to dream bigger than our backyards."

Annie nodded, her voice firm with conviction. "Yes, because in the midst of all this noise, there's a space where we can dream, where we can test the edges of what's possible. And who knows? Maybe one of these dreams, fueled by our magick, will take us to those moons or beyond."

Magic

Magic, akin to intention, is a form of magick. However, not all magick qualifies as magical. Magic, spelled without the "k," physically manifests as the art of creating mirages — enchantment, deception, or fleeting otherworldliness. Magic is the art of conjuring surprise and mystery through sensory distortions, playing tricks on our perception, eyes, and senses. It produces illusions that either confuse our grasp of reality or appear to defy its constraints, often leading us into a realm of amusement, humility, and curiosity.

Magic is the crafting of illusions and demands cognitive engagement. The fascination with magic, or the diverse techniques magicians employ to deceive audiences, whether through sleight of hand, misdirection, or grand stage setups, relies on the ability to think and be astonished. Magick, on the other hand, can exist autonomously of cognitive processes, evident in phenomena like the phases of the moon or the dance of the wind. The aim of magic is to create or witness an illusion so compelling that it seems to challenge reality itself, sparking surprise and wonder.

Viewing magic through the lens of illusion design highlights the physical principles at play in magic tricks. There are concrete formulas for constructing compelling magic tricks, tapping into both the physical and psychological aspects of human perception. Magic can be experienced in response to natural phenomena like the moon illusion or through the deliberate artistry of stage magic found in theaters, amusement parks, festivals, movies, birthday celebrations, and art installations.

Magic, as the orchestration of illusions, holds a significant place in the human experience, particularly in those moments of unexpected wonder where our senses are captivated by surprise. This enchantment is deeply rooted in the neuroscience of perception, involving a complex interplay of cognitive functions. When encountering a masterfully executed illusion, our brain engages in rapid pattern recognition, momentarily suspending our understanding of reality, and triggering a cascade of neurotransmitters linked to pleasure and intrigue.

The interplay between surprise and deception, however, navigates a nuanced terrain. While well-crafted illusions can inspire joy and awe, ill-conceived or malicious deceptions might evoke negative reactions such as confusion or fear. Magic, as a branch of magick, operates within this spectrum, highlighting the dual nature of physical manifestation and psychological impact. It serves as a poignant reminder that within the realm of wonder, magic has the dual capacity to both enchant and mislead, to elevate or unsettle.

The Magician

The pursuit of joy in life benefits from both the study and practice of magick, and the active engagement with magic. Magic, in its most captivating form, is not just an act of deception; it's a sophisticated interplay of engineering, mathematics, and the profound understanding of human perception.

The magician is not merely a deceiver but an architect of experience, meticulously constructing illusions that blend the predictable with the surprising.

Every magic trick has a blueprint, a sequence of actions that, when executed with precision, create an illusion that seems to defy the laws of physics or probability.

Like a well-orchestrated symphony, timing in magic is everything. The magician controls the tempo of revelation and concealment to maximize the impact of the trick.

Humans are pattern seekers by nature. Magicians exploit this by presenting patterns that lead the audience to one conclusion, while the actual mechanics of the trick follow a different, hidden pattern.

Despite the ethereal quality of magic, its execution is rooted in the physical world. The magician must understand the limitations and capabilities of physical objects and human senses.

162

The magician's art is less about believing in the impossible and more about making the possible appear impossible. This involves:

Using principles of mechanics, optics, and psychology to construct illusions that deceive the senses without breaking the laws of nature.

Unlike magick, which might seek to transform or transcend, magic revels in mystery for its own sake. It's about the joy of the inexplicable within the realm of the explainable.

Magic can transform the mundane into the extraordinary, showing that even in the most ordinary settings, there is room for enchantment.

The pleasure derived from magic lies in the experience of being baffled, in the art of being fooled in a way that sparks wonder. It's about the celebration of ignorance, not out of a lack of knowledge, but from the delight of being led astray in a controlled, skilled manner. Magic, through its mastery of physicalism, teaches us to find magic in the mechanics of our world, to appreciate the beauty of what is just beyond the obvious, and to enjoy the mystery crafted from the very fabric of reality.

History of Magic

Magic has historically been perceived across numerous cultures as a conduit for communing with deities or supernatural entities, seeking favor in the afterlife or protection in the present. Magicians were often regarded as intermediaries between the mundane and the mystical, with magic deeply interwoven with religious practices. Throughout history, magic has manifested in diverse forms, from West African Vodou to African American Hoodoo, and from ancient to modern practices like Paganism and Wicca. Many magical traditions, such as Santería and Kabbalistic magic, blend elements from multiple religious heritages into unique spiritual frameworks.

Nature-centric religions like Shamanism and Druidry incorporate indigenous magic, while Tantra and Taoism use magic alongside meditation for achieving harmony. Hermeticism and Thelema develop magical systems from esoteric philosophies, and Enochian magic, a relatively modern form introduced in the 16th century by John Dee, was believed to involve communication with angels through a divine language. Despite the variety in practice, religious magic commonly aims to forge connections with the divine, gain spiritual insights, and enhance personal power.

The transition of magic from a spiritual practice to a form of theatrical entertainment was influenced by several key factors, including the Enlightenment, which emphasized science and reason, favoring

164

skepticism that prompted magicians to incorporate scientific principles and psychological insights into their acts; urbanization and the rise of the middle class, which created a demand for novel forms of entertainment, leading magicians to refine their art and transform sleight of hand and illusions into a sophisticated performance art; and the growing demand for amusement as societies industrialized, causing magic to evolve into a popular form of spectacle where magicians became showmen who dazzled audiences with elaborate stage setups and complex tricks.

Magic's history is rich with luminaries who have each contributed uniquely to the art of illusion:

Harry Houdini transformed magic with his escape acts, emphasizing physical prowess and showmanship.

Criss Angel brings a modern, edgy mystique to magic, often blending it with street performance and extreme stunts.

Penn and Teller offer a blend of skepticism, humor, and revelation in their acts.

David Blaine redefines magic by performing in everyday settings, making the impossible seem possible in the open air.

Piff the Magic Dragon infuses magic with comedy, creating a whimsical, memorable experience.

Magic extends beyond mere entertainment into specialized areas like "gospel magic," where illusions serve to convey religious messages. Street magic also offers a unique angle, turning ordinary urban encounters into moments of wonder, providing both a source of income and an element of surprise in daily life, enriching the mundane with the extraordinary.

Natural Magic

In "The Magic of Reality," Richard Dawkins, a preeminent evolutionary biologist, champions the idea of natural enchantment, suggesting that the wonder typically associated with mystical or supernatural phenomena can be equally, if not more, found in the natural world and scientific exploration. Dawkins asserts that science, though often perceived as complex, offers a pathway to understanding and appreciating the marvels of existence, whether it's the inner workings of technology, the physics behind rocketry, the art of stage magic, or the biological processes that drive evolution and human physiology.

Experiencing true natural enchantment involves a combination of surprise and deep curiosity. This experience spans a wide range of emotions: like awe, amazement, intrigue, wonder, and delightful uncertainty. This sense of enchantment can be triggered by **human ingenuity like in m**agic tricks where sleight of hand and clever misdirection lead us to question what we see. **Psychoactive experiences are prevalent in m**oments where our perception shifts due to substances, trauma, or profound mental experiences, opening up new ways of viewing reality. Nature itself is an artist of the highest caliber, effortlessly weaving illusions like mirages in the desert where water appears where none exists, or the moon illusion where the moon looks larger near the horizon. Cliffs that seem to defy gravitational laws, or the intricate patterns of light and shadow on a forest floor that can conjure up images or guide one's path.

These natural phenomena serve as living evidence of how easily our perceptions can be influenced by the environment. The interplay of light, the structure of landscapes, and the natural world's ability to create visual tricks are not just beautiful; they are profound lessons in how we interpret our surroundings, reminding us that the magic of reality is all around us, waiting to be discovered through the lens of science and an open, curious mind.

Exercise #18: Cultivating a Meditation on Natural Magic

Richard Dawkins' concept of natural magic invites us to explore the wonder and enchantment of the natural world through a scientific lens. This exercise helps cultivate a sense of awe and appreciation for the natural world by engaging in mindful meditation and observation of natural processes.

Choose a peaceful location in nature, such as a park, garden, beach, or forest. Sit or lie down comfortably, so you won't be disturbed during meditation.

Close your eyes and take a few deep, calming breaths. Clear your mind of distractions and focus on the present moment.

Slowly open your eyes and begin to observe your surroundings. Engage all your senses: notice the sights, sounds, smells, textures, and even tastes if safe and appropriate.

Focus on a specific natural element or processes around you, like the movement of leaves in the wind, the flow of a river, or the growth of plants. Observe these elements closely, paying attention to its most intricate details.

As you observe, reflect on the scientific principles behind what you're witnessing. How does physics, chemistry, or biology explain this natural phenomenon? Imagine the delicate interactions of molecules, atoms, and forces that make this phenomenon possible.

Take a few moments to contemplate the sheer complexity and beauty of the natural world. Consider the eons of evolution and the interconnectedness of life. Reflect on the fact that this beauty and complexity emerged from natural processes, without the need for supernatural explanations.

In your journal, write down your thoughts and feelings about what you've observed and contemplated. Express gratitude for the opportunity to witness and understand the natural magic around you.

Commit to regular sessions of this meditation, ideally in different natural settings. Over time, expand your observations to encompass a wide range of natural phenomena, from the smallest insects to the grandeur of the cosmos.

Emergence

Emergence occurs when complex systems or processes manifest unexpected properties or behaviors that cannot be inferred simply from the sum of their parts. This phenomenon can be likened to natural magic, where the outcomes captivate and baffle, appearing illogical or even magical, yet they are the product of natural interactions and their intricate complexities.

Flocking Behaviors: The synchronized flight of bird flocks or the schooling of fish epitomize emergence. These collective behaviors give the impression of a choreographed dance or a single guiding intellect, yet each bird or fish responds solely to local stimuli and the movements of neighbors. This creates a mesmerizing spectacle that seems directed by an invisible hand, a grand conductor in nature, but it's really an emergent property of individual actions.

Pareidolia: Seeing patterns, like faces in clouds, is another example of emergence where our brains, from the simplest visual cues, construct complex images, illustrating how emergent phenomena can create enchanting illusions from the mundane.

Color Theory: When colors are combined, they can evoke emotions or atmospheres that surpass their individual effects. This emergent property in art and visual design can create a mood or ambiance that feels almost magical, a sum greater than its parts.

171

The Circus Atmosphere: Consider the circus with its vivid tent, eclectic mix of smells and sounds, the cacophony of music, and dazzling performances. Each element on its own is simple, but together they emerge into an immersive experience of wonder, transporting attendees into a realm of fantasy and excitement. The circus encapsulates the essence of emergent enchantment; the whole experience is far more captivating than any single act or smell or color could achieve alone.

Emergence, in this sense, can be seen as the physics of magic. It is characterized by several key aspects. These include individually simple yet collectively complex elements where small, simple interactions or elements can lead to complex and enchanting outcomes. This mirrors the principles of magic in which simple tricks, when combined, create astonishing illusions. Another aspect is surprise and wonder. Emergence often leads to surprise because the results are greater or different than what one might predict, fostering a sense of wonder akin to magic. Pattern recognition is also key. Our brains are wired to find patterns, and emergent phenomena exploit this by creating patterns that only become visible when viewed as a whole, not in their individual parts. Finally, there is the illusion of control. In both nature and magic, emergence can give the impression of a guiding force or intelligence, where none exists in a traditional sense. It represents instead the self-organization of simple elements under certain conditions. In essence, emergence is the unseen hand that orchestrates the symphony of the physical world, crafting what feels like magic from the chemistry of interaction, the physics of systems, and the mathematics of patterns, revealing that the enchanting aspects of life are often woven from the threads of everyday existence.

Synchronicities

Carl Jung's term synchronicity refers to seemingly acausal, yet profoundly meaningful coincidences. While often perceived as having a divine or esoteric origin, a physicalist interpretation suggests that:

Synchronicities might lack apparent causality, but they could stem from intricate, subtle links within the natural world. These connections can be so complex that they give the illusion of being orchestrated by a higher power.

Even when synchronicities feel divine or orchestrated, they often have a basis in the physical world. The complexity of these interactions can make the causality difficult to discern, leading to the perception of magic or divine intervention.

Synchronicities can be an emergent property of the universe where individual actions or events, when combined in a certain way, produce outcomes that appear to defy explanation but are, in fact, the result of natural processes.

Human brains are predisposed to find meaning and patterns. When we encounter synchronicities, our cognitive bias towards meaning-making can lead us to perceive these events as more than mere coincidence, sometimes attributing them to a divine or supernatural cause.

The feeling of a designed universe or a divine plan behind synchronicities can be understood as an illusion created by our interpretation of complex, chaotic systems. These systems, governed by physical laws, can produce outcomes that feel personal or purposeful.

When synchronicities occur, the conflict between the observable physical world and the sense of something beyond it can lead us to ascribe these events to a mystical or divine source. However, a closer examination might reveal underlying physical or psychological mechanisms at play.

By recognizing the physical causality behind synchronicities:

We demystify them without diminishing their wonder. Understanding that they might be rooted in the physics of complex systems does not strip them of their significance but instead reframes the narrative from the divine to the naturally profound.

We enhance our appreciation of the universe's intricate workings, where the ordinary can give rise to the extraordinary through the interplay of countless variables.

We invite scientific inquiry, encouraging us to look for the threads of causality that weave through what appears to be the fabric of fate or destiny.

Synchronicities, therefore, can be seen as a testament to the universe's complexity rather than evidence of the supernatural. They challenge us to expand our understanding of how the physical world can give rise to experiences that resonate on a deeply personal or spiritual level, all while adhering to the laws of physics.

The Science of Magic

Cognitive illusions are shaped by various physical states and processes. These include memory, where past experiences and stored information can influence how we perceive new stimuli, often leading to cognitive illusions. Motivations and emotions also play a role. Our emotional state or what we desire can color our perceptions, making us more susceptible to certain kinds of illusions.

The three fundamental pillars of magic, attention, expectation, and perception, are critical in constructing and experiencing illusions. Attention is a limited resource. It directs our focus to specific elements of our environment, inherently causing us to miss or ignore others. This selectivity is what magicians exploit to create their deceptions. Expectation is another pillar. When our anticipations are not met, it can lead to the perception of the impossible or surreal. Expectations shape what we notice, remember, and how we interpret what we see. Perception is the process by which we interpret sensory information. It is here where illusions take root, as our brains can be easily tricked into filling in gaps or making assumptions based on incomplete data.

Viewing magic through the lens of physicalism involves several aspects. Sensory data encoding is key. Magic tricks often depend on the way sensory information is processed. The brain's encoding of this information can be manipulated to produce illusions. Perception development is another aspect. Illusions play with how perception is constructed, revealing the brain's shortcuts and assumptions in interpreting reality. Expectation formation and influence are also involved. The study of magic shows how our cognitive frameworks are built upon expectations, which can be subverted to create magical effects. Cognitive biases are leveraged as well. Magic uses these biases, such as confirmation bias or the illusion of control, to enhance the impact of tricks. Attentional mechanisms are demonstrated too. Magic tricks are a practical demonstration of how attention can be directed or misdirected, showing the brain's limitations and capabilities in focus and awareness.

The scientific inquiry into magic illuminates information processing. It sheds light on how we process sensory inputs, make decisions, and form memories, often under conditions of deception or surprise. It enhances understanding of human perception. By understanding how illusions work, we gain insight into the human brain's ability to perceive and reinterpret reality, even when faced with misleading cues. It offers practical applications. Insights from magic can be applied to areas like psychology, education, and even the design of user interfaces where understanding perception is key. It promotes cognitive flexibility. Engaging with magic can encourage a playful questioning of reality, fostering cognitive flexibility and an openness to alternative explanations of sensory experiences.

From a physicalist perspective, magic is not just trickery but a profound exploration of how our brains construct reality from the physical inputs we receive. Appreciating magic can thus be beneficial, enhancing our understanding of cognition, enriching our mental lives, and prompting us to marvel at the complexity of our perception machinery.

Illusions

The multitude of tricks and their underlying mechanics create an unlimited canvas of wonder within the captivating realm of illusions and varied magic genres. From sleight of hand to mentalism, the possibilities are as vast as your imagination. While watching performances, navigate the nuanced distinction between magic and magick, carefully distinguishing illusions from the procedures. This investigation goes beyond the stage to the emergences and synchronicities of nature, broadcasting the universe's exquisite performance. Magic reminds us that the enchantment we seek is real despite deceptive.

Visual

The Necker Cube

A wireframe cube that appears to switch between two perspectives is the subject of this optical illusion. Despite the fact that the image is only 2D, our brains interpret it as a 3D entity that can be viewed from many perspectives.

The Hollow Face

A concave mask that appears to be convex from a particular angle is the subject of "The Hollow Face." Our minds fill in the blanks and interpret the mask as a convex face since they are accustomed to seeing convex faces.

The Blivet

This illusion, also referred to as the "impossible fork," involves a fork-like device that appears to have three prongs on one end and two on the other, despite the fact that it is physically impossible to manufacture such an object.

The Motion Aftereffect

After seeing a moving stimulus, the illusion involves a stationary item that appears to be moving. A transient adaptation that results in a perceived motion in the opposite direction is brought on by overstimulation of neurons that respond to motion direction.

Adelson's Checker-Shadow

This is an optical illusion that illustrates how the visual system is affected by context. In this illusion, a pattern of alternately dark and light squares surrounds a gray checkerboard. Some of the gray squares seem lighter or darker than others because of how the brain interprets visual information. Particularly, the dark squares around the gray squares make them appear brighter than the light squares around the gray squares. The gray squares are all the same shade of gray in reality, but because of the contrast with their surroundings, they appear to be distinct.

The Thatcher

When viewed upside-down, the reversed image of a face looks twisted and unnatural, but when viewed right-side-up, it appears normal.

Our brains analyze faces differently than other objects, and the inverted image interferes with how our brains interpret faces.

Auditory

Risset's Rhythmic Effect (Shepard tone)

An auditory illusion known as the Shepard tone or Risset's rhythmic effect causes a tone to appear to be constantly rising or falling in pitch while actually being at a fixed pitch. In order to get this effect, many pure tones that are spaced apart by octaves are superimposed. It appears as though the pitch is rising or falling continuously because each tone is progressively faded in and out as it is played. In order to convey a feeling of tension, suspense, or endless motion, the Shepard tone is frequently employed in music and sound design.

McGurk Effect

The McGurk effect is an audio-visual illusion that happens when the auditory and visual components of two different sounds are combined, giving the impression of a third sound. The spectator might hear the sound "da" if, for instance, a video of someone saying "ga" is shown while someone else is speaking "ba" on the audio track. This illusion results from the brain's integration of auditory and visual information, which can occasionally cause perception problems.

Speech-to-Song

A phenomenon known as the speech-to-song illusion occurs when a spoken sentence is repeatedly repeated to a listener and finally begins to seem more like a song or a musical melody than actual speech.

Speech Shadowing "Cocktail Party Effect"

The "cocktail party effect" is a type of auditory delusion that can occasionally be produced by speech shadowing. The ability to selectively attend to one auditory signal, such as a conversation with a particular individual, while ignoring other competing noises, such as background noise or other conversations, causes this effect. The spatial separation between the target and competing noises, as well as other elements like familiarity or emotional relevance, can enhance the cocktail party effect.

Tactile

Rubber Hand

The Rubber Hand Illusion is a tactile illusion that tricks your brain into thinking that a rubber hand is part of your own body. The participant's real hand is hidden from their view, and a rubber hand is placed in its position. Both the rubber hand and the real hand are touched simultaneously, and the brain begins to associate the sensation with the rubber hand, creating the illusion that the rubber hand belongs to the participant's body.

Thermal Grill

The Thermal Grill Illusion is an unusual tactile illusion that causes the sensation of extreme heat or cold, even when the temperature is moderate. It is created by placing two rows of warm and cool bars in contact with each other. When the bars are touched simultaneously, the brain is unable to distinguish between the warmth and coldness, resulting in the sensation of intense heat.

Cutaneous Rabbit

The Cutaneous Rabbit Illusion is a tactile illusion in which a series of taps in different locations on the skin creates the sensation of a rabbit hopping from one location to another. The illusion is created by the brain's inability to detect gaps in sensory input, resulting in a continuous sensation that seems to move from one point to another.

Phantom Limb

The Phantom Limb Illusion is a tactile illusion in which an amputee continues to feel sensations in a missing limb. The brain creates the sensation of touch, even though the limb is no longer there, leading to the feeling that the missing limb is still present.

Kinesthetic

The Kinesthetic Illusion is a tactile illusion that can create the sensation of movement in stationary objects. It is created by stimulating the muscles and tendons associated with the movement of the object, tricking the brain into thinking that the object is actually moving.

Cognitive

Anchoring Effect

This cognitive fallacy explains how people frequently base their decisions on the first piece of information they learn. When a car salesperson tells you, for instance, that a car costs $50,000, it could be tough for you to contemplate a lower price, even if it is more reasonable.

Framing Effect

When people respond differently to the same information depending on how it is given, a cognitive illusion known as the framing effect can transpire. Even though both statements fundamentally mean the same thing, individuals might be more likely to support a program if it is said to have a 90% success rate than if it is said to have a 10% failure rate.

Illusory Superiority

The propensity for people to overestimate their skills or performance in comparison to others is referred to as illusory superiority. People may rate themselves as better drivers than average, for instance, even though the majority of people cannot share this opinion.

Confirmation Bias

When people dismiss information that contradicts their preexisting beliefs or expectations in favor of information that supports them, they fall victim to this cognitive illusion. For instance, information concerning how vaccines damage people may stick in their minds more so than stories about how they prevent disease.

False Consensus Effect

The propensity for people to exaggerate the extent to which others share their thoughts or beliefs is referred to as the false consensus effect. People might believe that because they hold a particular political or other opinion, the majority of people must hold the same view.

Placebo Effect

When patients notice a positive improvement in their symptoms or condition following treatment that contains no active ingredients or therapeutic impact, they are experiencing the placebo effect. This may happen merely because patients think their treatment benefits them.

Halo Effect

The halo effect is a cognitive bias that occurs when our overall perspective of something influences our perception of its specific characteristics. It happens when we allow one good or negative trait to overwhelm our evaluation of unrelated traits. Based on inadequate knowledge, this bias can cause us to make assumptions or overgeneralizations. For example, if we find someone visually appealing, we may instinctively infer they have other desirable characteristics.

186

Similarly, if we have an unfavorable view of a brand, our perception of its many characteristics may be influenced.

CHAPTER 6

THE OCCULTIST

Lucian leaned heavily on his porch railing, the misty Appalachian morning stretching endlessly before him. His back ached, and his vision had been clouded for days, but he chalked it up to getting older. Annie, sipping coffee beside him, wasn't convinced.

"It's not just age, Lucian," she said gently. "The symptoms you're describing — the stomach pain, the nausea, the blurry vision — they're too interconnected to be random. Something's wrong."

Lucian grumbled but didn't argue. The truth was, he felt worse than he let on. His memory had started to slip, and the once-vivid colors of his surroundings seemed dimmer. Annie was right, but seeing a doctor meant confronting the insurance system, something he had little patience for. Appointments took weeks, sometimes months, and the costs were crushing.

Annie, as always, had another plan. "I've got something that can help. It's not conventional, but it'll get us answers faster than waiting on an insurance approval."

Lucian narrowed his eyes. "What do you mean?"

Annie's lab, hidden in the modest shed behind her cabin, was a marvel of ingenuity and resourcefulness. Lucian had always assumed it was just for her "experiments," but today, she revealed its full scope. Among the clutter of beakers, screens, and what looked like a mechanical octopus, sat a sleek device she called a chemputer. Nearby, a portable MRI machine whirred to life.

"How in the world did you get this?" Lucian asked, astonished.

Annie shrugged. "I've been building up my lab piece by piece for years. Most of this tech is repurposed or open-source. The government doesn't make it easy—everything's taxed or regulated to death—but I'm working on ways to make these resources more accessible. Technology like this could save lives if it weren't so tied up in bureaucracy."

Lucian hesitated. "So, this machine is gonna... see inside me?"

Annie nodded. "That's the idea. If there's something compressing your nerves or affecting your organs, this will show us."

Despite his skepticism, Lucian agreed. The scan took minutes, and the results were immediate. Annie pointed to the screen, where a bright cluster of compressed nerves near Lucian's spine glowed.

"There it is," she said. "It's causing all those symptoms. If we address this, you'll start to feel a lot better."

Lucian stared at the image, the weight of his ignorance sinking in. "I never would've guessed. I thought it was just... everything falling apart."

"That's the thing about the body," Annie said. "So much of it is hidden, just like the world around us. Without the right tools or knowledge, we're blind to it."

190

As Lucian recovered in the following weeks, he found himself thinking more and more about Annie's lab and the technology she wielded so deftly. He'd always been a humble man, content with his limited understanding of the world. But now, he felt something shift.

One morning, he sat on his porch, watching the sunrise. The colors seemed a little brighter, his vision clearer than it had been in years. He thought about the boundaries of knowledge—how every person, no matter how wise, eventually hits a wall where their understanding ends. He thought about microscopes, chemputers, and Annie's "occult" machines that revealed the unseen.

Occult. The word took on a new meaning. It wasn't about secrecy or mysticism; it was about uncovering what was hidden, seeking out the invisible threads that connect everything. He remembered Annie's explanation of occult blood tests, where doctors use microscopes to analyze particles too small for the naked eye. Without those tools, even the smartest people would miss critical details.

For the first time in years, Lucian felt a spark of curiosity. What else was out there, waiting to be discovered? What other assumptions had he made, only to find they were wrong?

He thought about the "fools" in Annie's network—the goths, punks, and misfits who the world dismissed as dreamers or troublemakers. Maybe they weren't fools at all. Maybe their curiosity, their refusal to accept the limits placed on them, was the smartest thing of all.

When Annie joined him on the porch that morning, Lucian smiled. "I've been thinking about what you said... about knowledge and boundaries."

"Oh?" Annie asked, curious.

"There's joy in not knowing everything," Lucian said. "In realizing how much more there is to learn. Even stage magic... it's all about what you can't see. People like surprises. They like the wonder of the unknown."

Annie nodded. "That's what keeps us going, isn't it? The curiosity, the challenge, the chance to discover something new."

Lucian leaned back in his chair, the aches in his body fading into the background. For the first time in weeks, he felt alive. There was so much he didn't know—but instead of fear, he felt hope.

"Let's see what else is out there," he said, raising his coffee mug. Annie raised hers too, their laughter mingling with the sound of the wind in the trees.

Occultism

The word "occult" comes from Latin, like "oculus" for eye and "occultare" for hiding. It's about diving into the mysteries of life, where ignorance meets the thirst for knowledge. Sure, there's some crossover with cults, especially when people get into the weird or dark stuff, but they're not the same thing. Think of occultism as a quest for knowledge, not some cult gathering. Cults can latch onto anything, including occult ideas, to seem mysterious or powerful.

Occultism is all about embracing the unknown, exploring beyond what we know. There's a whole universe out there we haven't even scratched the surface of, from new science to undiscovered worlds.

The heart of occultism is about seeing what's hidden, craving discovery at the edge of existence. It's about dealing with the uncertainty life throws at us—whether it's the fear of a car crash, getting cancer, or a rocket blowing up.

Other terms might capture this dance with the unknown, but "occultism" is about living with eyes wide open, seeking knowledge amidst the risks and mysteries of life. It's about brave exploration into new thoughts and experiences, not just dark intentions but also the wonder scientists chase in understanding the natural "magic" of the world.

The four pillars of occultism are:

Ignorance - We start not knowing anything.

Awareness - We learn, we grow, we see more.

Inquiry - We keep asking questions, driven by curiosity.

Prediction - We make guesses, plan for what might come next.

In this light, occultism is a journey through cycles of ignorance, enlightenment, constant questioning, and speculative prediction. It's about embracing the unknown, expanding our consciousness, and daring to predict based on what we've learned.

The Occultist

Embracing the essence of occultism means boldly stepping into unknown realms, not out of reluctance to learn, but from an understanding that growth often leads to uncharted territories. In spiritual traditions, magick and occultism coexist in a symbiotic dance, where occultism probes into the shadows, misconceptions, and missing pieces, while magick focuses on tangible designs and rituals. An occultist navigates the labyrinth of danger, uncertainty, limits, paradoxes, and loss with a nuanced understanding.

The occultist is one who gazes unflinchingly into the void, driven by a complex thirst for knowledge. This desire draws individuals bravely into the abyss of black holes and the trials of shattered spirits, pushing them ever forward. At its core, occultism champions the embrace of ignorance, learning, and overcoming challenges. The occultist acknowledges their own folly, driven by an unquenchable quest for knowledge, adaptable to new ways of solving problems, dreaming, and experiencing life.

The Four Pillars of Occultism:

Ignorance - This pillar acknowledges our inherent limitations, echoing the agnostic stance that there's always more to learn. In occultism, ignorance is not a barrier but an expansive field for growth, inviting inquiry and unexpected learning, turning the unknown into a playground for exploration.

Awareness - Here, we delve into the ability to form mental images of our existence from sensory data, through our evolved bodies. Awareness is about the vivid connections we make with the world—be it the beauty of nature, the scent of the air, or the melody of music. It's the "oculus" of occultism, where seeing equates to understanding.

Inquiry - This pillar opens the door to discovery and the conscious expansion of thought. Inquiry thrives on questioning and oracular practices, engaging with awareness in a dance of curiosity, analysis, and creative exploration. It's a never-ending loop that not only deepens understanding but also reshapes our mental landscapes, linking ignorance with knowledge in a dynamic cycle.

Prediction - The final pillar uses accumulated knowledge to forecast future scenarios, from weather patterns to market trends or personal life paths. In high-stakes fields like science, philosophy, and business, prediction is deeply intertwined with occultism. It's about making guesses from a place of uncertainty, guiding us through the vast unknown with curiosity as our compass.

Occultism isn't just about the mystical or the hidden. The core of occultism is a philosophy of engaging with life's mysteries, embracing our limitations, and expanding our consciousness through the interplay of ignorance, awareness, inquiry, and prediction. It's a journey of transformation, encouraging us to look beyond what's known and to live with an adventurous spirit in the face of life's great uncertainties.

196

TSOHPM

Exercise #19: Summon the Fool

To critically analyze your own moments of stupidity, explore the fine line between beneficial ignorance and harmful folly, questioning if a dash of stupidity is healthy, and debating the pursuit of all-knowingness.

Choose a spot where you can think deeply - maybe a quiet room or a bench in a park. No need for decorations, but if you're inclined, place something that represents folly, like a comic book or a puzzle with missing pieces, to remind you of the exercise's theme.

Sit down, close your eyes, and imagine sitting across from your own 'Fool' self - the part of you that acts or thinks in ways you might later call stupid or naive.

Advantages of Stupidity:

Recall a time when not knowing something led you to a creative solution or an unexpected discovery.

When has being wrong or ignorant actually propelled you to learn more or grow?

Think about how not knowing everything can be liberating, allowing you to enjoy life without the burden of omniscience.

Disadvantages of Stupidity:

Reflect on when your ignorance led to mistakes, possibly even harm or loss.

198

Consider how not knowing can slow you down, make you miss opportunities, or lead to poor decisions.

Think about instances where your lack of knowledge left you vulnerable to being misled or exploited.

Now, debate with yourself:

Is Stupidity Sometimes Good? Argue whether a little ignorance or foolishness can make life more enjoyable, less predictable, or even more human.

Discuss the implications of striving for omniscience. Would it strip away the joy of learning, the humility of not knowing, or even the fun of making mistakes?

Write down or voice three key takeaways:

1. An instance where your stupidity was an asset.
2. A time when it was a clear liability.
3. Your opinion on whether embracing some folly can enrich life.

Thank the Fool for the lesson. Consider how acknowledging and occasionally embracing your stupidity might actually lead to wisdom, not just about yourself but about the human condition. How can you apply this understanding to navigate life more authentically, appreciating both the knowledge you seek and the ignorance you sometimes cherish?

JANGLED JESTER

Occult > Cult

Despite sharing a similar sound, occultism isn't the same as cultism. While occultism can occasionally intersect with cult activities, they are fundamentally distinct. Occultism is rooted in a philosophy of curious exploration and knowledge-seeking, whereas cultism often involves groups engaging in behaviors widely considered unacceptable. Occultism has been unfairly painted as sinister by religious figures and popular culture, but it's essential to look at its core principles: a philosophy driven by curiosity and the pursuit of hidden knowledge.

The human spirit naturally seeks to explore mysteries, embrace ignorance, and venture into uncharted territories throughout life. Even in medicine, terms like 'occult blood' refer to detecting hidden elements, guiding us towards understanding the unknown. Occultism, then, is an everyday yet inspiring philosophy that champions curiosity about our existence.

Despite its enigmatic nature, the pursuit of knowledge inherent in occultism has psychological benefits. We often encounter problems that seem unsolvable, only to turn these challenges into opportunities for remarkable discoveries.

At its core, occultism is an endless quest for new insights. It acknowledges our human limitations - our finite capabilities, resources, and opportunities - yet encourages us to push these boundaries with curiosity and a touch of folly.

Occultism is a tangible journey of exploration and discovery. It can be daunting or challenging, confronting us with our fears, limitations, and missed experiences. However, it promotes humility, respect for individuality, and a deeper dive into the unknown, examining the essence of our existence. There might be apprehension about engaging with occultism if one doesn't value seeking new wisdom, but it embodies a spirit of eagerness, passion, and interest.

On the flip side, cults are characterized by dogma and extremism, often employing tactics that society broadly rejects. Cult behavior aims to isolate individuals from external influences, leading to a tunnel vision where their beliefs are seen as superior and all-encompassing. As Richard Grannon describes, this can manifest in a "cult of one," where an individual assumes absolute authority, demanding others follow without question.

Cult behavior typically suppresses questioning, critical thinking, and open dialogue. Emotional manipulation, through guilt, fear, or affection, is a hallmark, alongside isolation from diverse perspectives, further entrenching rigid beliefs. The impact of such dynamics, while concentrated within a group or individual, can extend far beyond, affecting communities and individuals around them.

While there's historical overlap where occult practices have been part of cult rituals, it's vital to differentiate occultism's foundational philosophy of curiosity and knowledge-seeking from the control and isolation of cults. The word "occult" simply means "hidden" or "esoteric" with no inherent negative connotation. Instead of focusing on the aberrations, we should highlight occultism's aim of exploration and discovery. By connecting occult practices to positive, societal, and enriching pursuits, individuals can steer their journey of inquiry towards beneficial outcomes, fostering rational predictions, and advancing into the unknown with intent and purpose.

Secular Occultism

Occultism transcends traditional boundaries of religion or secularism. It can be embraced by those who follow a specific faith, those on an agnostic path, or even those with atheistic perspectives. Occultism harmonizes with both scientific inquiry and mystical exploration. When viewed as a philosophy of curiosity-driven knowledge-seeking, it can profoundly influence many facets of human life, fostering a reverence for ignorance, awareness, inquiry, and the art of prediction. It encourages us to courageously delve into the unknown, equipped with tools and strategies that lead us into new realms of magick and occultism.

Secular occultism represents a worldview where the journey is an autonomous adventure into the unexplored. At its heart, it champions personal experimentation, allowing individuals to investigate life's mysteries on their own terms. There's an acknowledgment of the value of voluntary association within this sphere, where people might choose to come together to share knowledge, experiences, or insights of their own free will. However, the essence of this approach remains one of independence, urging individuals to carve their unique paths through the vast, enigmatic landscape of existence. This perspective prioritizes personal agency and autonomy, empowering people to craft their narratives, explore mysteries that capture their imagination, and make predictions aligned with their personal understanding of the cosmos.

Curiosity

JJ's top recommended metaphysics shop, Dilly's, is tucked away in Kingsport, Tennessee, and is now run by the daughter and son-in-law of its founders, Dillard and Sally Clark. Dilly's is a treasure trove filled with candles, crystals, metaphysical and occult literature, local artwork, spell oils, herbs, ritual tools, tapestries, and tarot decks. It's a haven for those curious about the esoteric, offering an open-minded environment. The bookstore section doesn't shy away from diversity, housing atheist and scientific readings alongside texts from every major religion.

Occultism embodies the perpetual cycle of curiosity, moving from ignorance through awareness, into inquiry, and onward to theory and prediction. Curiosity is the thread that weaves through the pillars of occultism, acting as the catalyst that drives individuals along this transformative path. The occultist embraces the active pursuit of knowledge, venturing into realms of awareness where they confront both their limitations and the boundless possibilities beyond. This ongoing cycle ensures that the seeker is never trapped in a mire of ignorance but instead continuously evolves, questions, and refines their understanding to navigate the complex quilt of reality. In this way, curiosity lights the path of occultism.

Exercise #20: Metaphysical Exploration and Community Connection Challenge

Take a trip to a local metaphysical shop. As you approach and enter, be mindful of your surroundings. Observe the displays of crystals, the variety of tarot decks, the shelves of mystical books, and any unique artifacts. Let your senses guide you.

Bring a notebook or use your phone to take notes. List the items that resonate with you, noting the energies or feelings they evoke. Write down any questions or observations that come to mind while you explore. This isn't just about cataloging; it's about connecting with the energy of the space and its content.

Initiate conversations with the staff or fellow visitors. Ask about the significance of certain items, seek recommendations tailored to your interests, and share your curiosity about metaphysical practices. This interaction is key to gaining deeper insights and fostering a sense of community within this metaphysical haven.

Look for flyers, notice boards, or digital screens advertising local metaphysical events, workshops, or gatherings. Make note of any that intrigue you, like classes, meditation groups, or public rituals. These events are your gateway to expanding your knowledge, networking with like-minded individuals, and actively engaging in the metaphysical community.

After your visit, set aside time for reflection. Revisit your notes, focusing on the items that particularly moved you and the nuggets of wisdom from your conversations. Research the events you've noted down, considering how you might participate. Think about how you can weave new metaphysical ideas or practices into your daily life. Embrace the ongoing journey of inquiry and the community spirit you've encountered, using this experience as a springboard for personal growth and connection.

CHAPTER 7

THE SCIENTIST

Annie's love for science grew from a mix of childhood curiosity and the experiences she gained while traveling through Appalachia. Moving to the mountains for her research, she found herself drawn to the resourcefulness and practicality of the locals, especially Lucian. A lifelong Appalachian, Lucian wasn't a scientist by trade — he was a retired mechanic and farmer — but Annie often said he embodied the scientific spirit. "Appalachians are incredible stewards of the Earth," she told him one evening. "Even with their love of old traditions, they're practical and innovative when it comes to preserving the land."

Lucian chuckled, shaking his head. "Don't know about all that, Annie. I just do what needs doing."

But Annie saw more. She admired how Lucian could fix just about anything with spare parts and ingenuity, how he knew the rhythms of the land as well as any ecologist. His deep respect for nature reminded her why she had chosen science in the first place — to understand and care for the world, even if it often felt like an uphill battle.

Science, Annie explained to Lucian, wasn't all discovery and progress. "It's hard," she said. "Sometimes I feel like a failure. Like I'm stuck at the edge of my own stupidity." She told him about the tidal waves that had devastated North Carolina after Hurricane Helene. "We had all the data, all the models, and we still didn't predict it right. It's frustrating. There's so much we don't know about the oceans, the universe, everything."

Lucian nodded thoughtfully. "Sounds like farming," he said. "You can do everything right, but there's always a storm you didn't see coming."

That comparison stayed with Annie. In many ways, science was like farming, a constant process of trial and error, planting ideas and nurturing them, knowing some would fail but hoping others would grow. Still, she struggled with the loneliness of being a scientist in a region where faith and tradition often took precedence over critical inquiry. She respected the religious communities of Appalachia but wished there were more people like her who are curious, skeptical, and enchanted by nature for its own sake.

"It's not that I think faith is bad," she told Lucian. "It's just... I wish more people could see the magic in science. The way physics and chemistry create all this beauty without needing anything supernatural."

Lucian smiled, remembering their shared experience at the technology magic show in the mountains. "Yeah, like that show we saw. Physicalist magic and using the wonder of science to inspire people. It's like alchemy, but without the mysticism. I want people to see chemistry as enchanting as any spell."

She started blending scientific demonstrations with a sense of wonder, using her research to connect with the community. "It's about making people curious," she said. "Helping them see that the natural world is incredible enough on its own. You don't need spirits or gods to be amazed."

Lucian found himself drawn into Annie's world. He'd always thought of science as something for people in labs, not someone like him. But Annie showed him how science was already a part of his life. In the way he fixed machinery, rotated crops, and read the weather. "Science isn't just for folks in white coats," she told him. "It's for anyone who asks questions and looks for answers."

As he spent more time with Annie and her colleagues, Lucian began to imagine a new Appalachia. Some parts of the mountains remained untouched, with kudzu swallowing old electric poles and rusted cars. But other areas buzzed with drones, electric vehicles, and hidden data centers. The physicalist magickians, scientists like Annie, were at the forefront of this transformation. They developed medicine with fewer side effects, restored sight to the blind, and gave children dreams of space travel and augmented reality.

Lucian saw the parallels between their work and his own life. "Every day's an experiment," he said one evening, sitting on his porch. "You try something, see if it works, and if it doesn't, you try again."

Annie smiled. "Exactly. That's science."

For Lucian, it was a revelation. And for Annie, it was a reminder of why she loved Appalachia. It is the people, the resilience, and the ability to blend the old with the new. Together, they found common ground in their shared respect for the land and their hope for a better future. In that quiet corner of the mountains, they proved that science wasn't just about answers but about the process, the questions, and the connections made along the way.

Science

Through empirical observation, experimentation, and evidence-based reasoning, science stands as the forefront of understanding the natural world. The scientific method is considered the most reliable approach for acquiring knowledge, focusing on explanations framed within natural laws and physical processes. This perspective rigorously evaluates hypotheses, including those involving supernatural or metaphysical claims, but ultimately prioritizes tangible, observable phenomena in the pursuit of a comprehensive understanding of the universe based on verifiable, reproducible evidence.

Even individuals not formally engaged in scientific research can develop robust systems that enhance both their everyday and more extraordinary lives. Non-religious practitioners, for instance, can craft systems that are both dependable and highly effective. This is achievable by adopting scientific methodologies, which emphasize the collection of empirical data and the practical application of evidence-based reasoning. These principles allow for the creation of frameworks that are not only grounded in reality but also adaptable to various aspects of human experience, blending the mundane with the seemingly magical through a lens of scientific inquiry.

The Scientist

In essence, scientists value empirical observations and real-world experimentation, structuring their work around measurability through data collection and evidence-based assessments. This approach can be applied to evaluate the effects of rituals, spells, or sensory experiences, allowing all practitioners to enhance the reliability and efficacy of their occult and magickal systems by adopting a disciplined, test-driven methodology.

This integration of science into the realms of the occult and magick serves to bridge the gap between seemingly disparate worlds, showcasing how empirical research and evidence-based practices can enrich spiritual endeavors. By merging the spiritual with the systematic, practitioners can achieve new depths of insight and mastery in exploring the enigmatic aspects of existence. Science's influence extends to individuals of all beliefs, offering benefits that transcend religious boundaries.

Indeed, people with deep religious convictions have made significant scientific contributions, benefiting from advancements in medicine, technology, and other fields. Many doctors, engineers, and scientists hold sincere religious beliefs, illustrating that science and faith can coexist harmoniously. However, skepticism or fear of what scientific inquiries might reveal can lead to resistance or avoidance. Unethical scientific practices have at times caused social discord and infringed upon human rights. Thus, the rigorous nature of science requires democratic oversight to prevent its misuse for harmful or nefarious purposes, emphasizing the critical role of ethical governance and responsible application.

Moreover, science, magick, and the occult do not have to be at odds; they can complement each other. The agnostic principle that acknowledges our limited knowledge aligns well with both scientific inquiry and the natural aspects of occultism. By embracing both agnostic science and occultism, individuals gain a more profound understanding of the world, recognizing that while not everything can be known, the pursuit of knowledge remains a paramount human endeavor.

Science isn't an adversary to occultism; rather, it acts as a companion, providing methodologies, tools, and procedures for exploring the uncharted, gaining insights, and fostering innovation. It encourages us to study, explore, and innovate, contributing to human survival and progress.

The scientific method, deeply embedded in physicalism, agnosticism, and secular occultism, forms the basis for empirical investigation. It starts with meticulous observation, resonating with physicalism's focus on the tangible aspects of nature. Through hypothesis formation and controlled experiments, it reflects the agnostic acknowledgment of our incomplete knowledge. The process of analyzing data and drawing conclusions parallels secular occultism by promoting a systematic exploration of hidden truths.

To promote ethical practices, unbiased research, and equitable knowledge dissemination, science must be democratized. The parallels between science, magick, and occultism become clear, as all involve a quest for discovery and systematic exploration of the unknown. Through its democratic application, science becomes an enlightening force, safeguarding against abuse while offering a complementary path for occultists and magicians to uncover the unseen truths of reality. Together, they forge a harmonious partnership between the known and the unknown, guiding humanity toward a deeper understanding of the mysteries that envelop us.

Scientific Terms

Controlled experiments are used in science to test ideas and assess the efficacy of magickal techniques. To determine the efficacy of various techniques, this method entails defining precise variables and testing circumstances, collecting data methodically and objectively, and analyzing the outcomes. Instead of depending on anecdotal or subjective evidence, evidence-based procedures involve analyzing claims based on empirical evidence and objective information. Using the greatest available data, this strategy involves critically assessing the body of prior research to determine which techniques are most likely to be successful.

Observational studies involve keeping track of natural occurrences like animal behavior or the results of specific environmental conditions. Patterns and correlations that may be relevant in the construction of magickal systems can be found through this research. Utilizing mathematical models, statistical analysis examines data to determine the likelihood of various events. This method can aid in locating important patterns and connections in data that may not be immediately obvious.

Replication is a process of doing new experiments or research to verify the validity of past results. This strategy aids in ensuring an experiment's findings are valid and not the result of pure chance. Peer review entails reviewing and scoring research studies and experiments by other participants in the subject. By taking this technique, research is more likely to be of good quality and have reliable conclusions. Empirical research and scientific testing are useful tools for creating effective magickal systems or examining them, but they might not be applicable in all circumstances. In situations where the data is confusing or inconclusive, subjective experience or intuition may serve as better guides for making decisions.

While periodically inaccurate, **intuition** is a great asset for speedy decision-making, pattern detection, and creative thinking. When combined with skill and experience, it complements rational thought, increases empathy, and can lead to novel discoveries and solutions. Intuition works quickly and automatically, drawing on a variety of specialized elements. One important factor is the brain's intrinsic ability to recognize patterns. It quickly detects links between current conditions and memories of earlier experiences, enabling for quick judgments and decisions without much conscious effort. In addition, intuition is based on heuristics, which are cognitive shortcuts formed via frequent exposure to comparable situations.

These mental techniques provide obvious instructions to simplify decision-making processes. In addition, our emotions and physical signals, such as gut sentiments or physiological sensations, might influence our intuition. Such nonverbal messages are processed quickly by our subconscious brain, which contributes to the quick and automatic character of intuitive answers. Intuition is resourceful, but it is crucial to realize that it isn't perfect. It is subject to biases and flaws. As a result, it is necessary to complement intuition with critical thinking and factual data to ensure any decisions and judgements are founded on a thorough and well-informed approach. We can create more trustworthy and balanced assessments of situations and improve our overall decision-making process by combining intuition with intellectual analysis.

The Scientific Method

Start by pinpointing a specific question or problem you aim to explore or resolve. Be precise and detailed in how you frame your inquiry to ensure clarity.

Gather background information on your topic. This can involve reading literature, reviewing scholarly articles, or consulting with experts to gain a comprehensive understanding of the subject matter.

Formulate an educated prediction or hypothesis that offers a potential answer or explanation to your question. This hypothesis should be both testable and grounded in the research you've conducted.

Design an experiment or a series of tests to rigorously evaluate your hypothesis. Clearly define your variables, controls, and experimental conditions to ensure your results are measurable and replicable.

Execute your experiment, meticulously recording all data and observations. Ensure you collect enough data to allow for meaningful analysis.

Analyze the data using appropriate statistical methods if necessary. Look for patterns, trends, or significant relationships in your observations.

Based on your data analysis, conclude whether your hypothesis is supported or refuted. Maintain objectivity and impartiality in your interpretation.

Share your findings with others through scientific papers, presentations, or discussions. Clearly outline your methodology, results, and the implications of your conclusions.

Reflect on your results and their impact on your original question or problem. If necessary, revise your hypothesis and experimental design for further investigation.

Recognize that the scientific method is iterative. If your initial experiments don't provide clear answers, refine your approach and repeat the process to deepen your understanding.

Throughout, employ critical thinking. Challenge assumptions, consider alternative explanations, and be open to changing your perspective based on new evidence.

Adhere to ethical principles in all aspects of your research. Ensure the safety and well-being of all participants or subjects and comply with ethical guidelines relevant to your field.

Cultivate a sense of curiosity in everyday life. Engage with the world by asking questions, seeking answers, and approaching challenges with an inquisitive and exploratory mindset.

Do not be discouraged by failures or inconclusive outcomes. In science, such results often pave the way for new insights and revised theories.

Open Science

Open Science is a revolutionary movement within the scientific community that promotes transparency, collaboration, and accessibility. It was created not merely to push the boundaries of knowledge, but also to challenge and prevent academic elitism. In an age when remarkable claims can be used to boost authority, Open Science works to democratize the research process, guaranteeing that the benefits of scientific discovery are available to all civilization.

Academic elitism, or the belief that knowledge and authority are restricted to a small group of people in academia, has long been a source of worry in scientific circles. It can erect obstacles to knowledge and prevent underrepresented groups from participating. Open Science combats elitism by encouraging transparency in research methods and outcomes as well as interactions with common people. Researchers are urged to publicly disclose their data, methodology, and conclusions, allowing their work to be scrutinized and replicated. This open exchange of information breaks down the barriers that traditionally existed between academia and the general public, allowing individuals to directly engage with scientific research.

Transparency is a crucial ingredient of Open Science. It becomes easier to identify and remedy cases of scientific misconduct, fraud, or tyranny by opening up the research process. The scientific community's and the general public's joint scrutiny contributes to the integrity of research undertakings. Researchers are held accountable for their work, and assertions are rigorously scrutinized. This transparency not only improves the quality of scientific results, but it also prevents research from being used to exercise excessive authority or propagate repressive ideas.

Open Science enables diverse voices to actively participate in scientific conversation. It encourages people from all walks of life to interact with complicated scientific fields and participate to ongoing discussions. Citizen science projects, for example, allow non-experts to work alongside researchers, broadening the pool of knowledge producers. Open access papers and open-source software cut entry barriers even more, making scientific materials available to a global audience. This openness not only enriches the scientific community, but it also ensures that the benefits of research are distributed more equally.

By breaking down traditional academic silos, Open Science has the potential to stimulate innovation. When scientists from different fields work together freely, they can bring new views to challenging challenges. Open data efforts make it easier to share significant research assets, which speeds up the pace of discovery. Furthermore, by engaging the general public in scientific pursuits, Open Science taps into a larger pool of creativity and skill, stimulating innovation that would otherwise go unnoticed.

Open Science is a paradigm shift in how research is conducted and shared. It actively attempts to combat academic elitism and the use of science for oppressive reasons by fostering transparency, accountability, inclusivity, and innovation. It encourages people from all walks of life to participate in the pursuit of knowledge, ensuring that research serves the greater welfare of humanity rather than being restricted to the domain of exclusive institutions or commercial interests. Open Science is a potent tool for democratizing research and improving human growth, eventually confirming scientific inquiry's collaborative and humanist attitude.

Sci-Hub

Sci-Hub is a website that provides free access to millions of scientific articles that are normally hidden behind paywalls. It was founded in 2011 by Alexandra Elbakyan, a computer programmer from Kazakhstan, with the goal of removing barriers to knowledge and promoting open access to scientific research. Many scientific articles are published in academic journals that require a subscription or payment to access. This can be a significant barrier for people who cannot afford to pay for access or do not have institutional affiliations. Sci-Hub helps to bypass these paywalls by providing a database of millions of scientific articles that can be accessed for free.

To use Sci-Hub, users simply enter the URL or DOI of the article they want to access, and the website retrieves a copy of the article from its database and provides it to the user. The website has faced legal challenges and has been blocked in some countries, but it remains a popular resource for those seeking access to scientific knowledge.

The Scientific Renaissance

Many scientists are on a quest to dismantle the barriers that separate the public from the radical expansions in scientific discovery and innovation. While revolutionary research and transformative ideas are primarily developed in laboratories and academic institutions, there's a growing realization that these discoveries should also be accessible to a broad and diverse audience. Scientists are increasingly seeking public involvement, understanding that the participation of ordinary citizens is crucial for enhancing scientific literacy and addressing global challenges.

Historically, there has been a significant divide between the scientific community and the general public. Scientific literature, once the domain of scholars and researchers, remains largely inaccessible to laypersons. Complex jargon and technical language often serve as barriers, deterring many from engaging in scientific discourse. Moreover, the pace of scientific advancement has accelerated, making it challenging for individuals to keep up with the latest discoveries and achievements.

Recognizing the need for change, scientists and organizations like Jangled Jester are actively seeking ways to communicate with the public. We understand that involving the general public in the scientific process can have profound benefits. Public participation not only empowers individuals but also opens up new avenues for scientific exploration by fostering curiosity, enhancing communication, promoting critical thinking, and elevating scientific literacy.

227

One notable trend is the rise of citizen science projects. These initiatives encourage everyday people to engage in real scientific research. Whether it's counting birds in local parks, monitoring environmental changes, or contributing to medical research through online platforms, citizen scientists significantly contribute to the body of scientific knowledge. This collaborative approach not only accelerates data collection but also instills a sense of ownership and passion among community members.

Science communication is another vital strategy for engaging with the public. Scientists are leveraging social media, podcasts, videos, and public lectures to make their work more accessible. This makes science not only informative but also entertaining, with its inherent debates, mysteries, and dramatic discoveries. Science communicators translate complex concepts into digestible content, demystifying scientific processes and making them relatable to daily life.

The benefits of greater public interaction with science are numerous. It fosters a more scientifically literate and engaged society, capable of making informed decisions on critical issues like climate change, public health, and technology. It inspires people to pursue careers in STEM, ensuring a vibrant scientific community in the future. Additionally, public input can spark new research directions, bringing fresh perspectives and innovative solutions to persistent problems.

In an era where scientific development increasingly affects our lives, the necessity for public participation has never been greater. Scientists are stepping up, recognizing that collaboration with the general populace can unlock new dimensions of scientific understanding. As we move towards a more adventurous and creative scientific renaissance, the involvement of everyone, regardless of prior scientific experience, is essential to securing a bright and informed future.

Exercise #21: Interview a Scientist

Choose a scientist or researcher whose work or area of expertise captivates you. This could be a well-known figure in their field, a professor at a local university, or someone working at a nearby research institute.

Research the scientist's background, their field of study, and any recent publications or projects they've been involved in. This background knowledge will help you craft insightful interview questions.

Compile a list of questions you'd like to ask. Start with basic inquiries about their field, then delve into more specific topics that pique your interest. Consider asking about their motivations, the challenges they face, and the potential impact of their work.

Reach out to the scientist to request an interview. Communicate your genuine interest in their work and clearly state the purpose of the interview. Be polite and professional in your approach.

During the interview, respect the scientist's time by being concise and attentive. Begin by introducing yourself and expressing your interest in their work. Ask your prepared questions, but also be ready to follow up on their answers to explore topics further.

Post-interview, reflect on the knowledge and insights you've gained from the scientist's perspective. Consider how their research fits into broader scientific and societal contexts. Maintain your curiosity by continuing to explore scientific themes through additional reading, study, and engaging with other experts across various disciplines.

The Science of the Supernatural

Exploring the supernatural through science invites us to reconsider how we label the unknown, moving away from the idea that it defies physics to the notion that it involves physics that we have yet to fully understand or determine. Instead of dismissing paranormal phenomena outright, this perspective treats them as potential physical processes awaiting interpretation by contemporary science. Richard Dawkins might argue that "supernatural" is a misleading term, but advocates like Jangled Jester suggest it points to phenomena that lie beyond our current comprehension — physical in nature but superior to our knowledge or not yet fully explained.

Consider, for example, a quantum computer: to a cave person, its operation would appear supernatural because the underlying physical processes are not understood. However, the quantum computer operates within the laws of physics all along; it's just that those laws are not yet grasped by the observer. This approach to the supernatural mirrors the essence of occultism, where what seems mystical or magical might simply be the result of physics we haven't deciphered yet. This perspective underscores that the supernatural is not an exception to physical laws but rather a frontier where our understanding of physics is incomplete. It echoes the core occult belief that human knowledge is an ongoing journey, reminding us that there is always more to learn — aligning with the inquisitive spirit of occult practices. Thus, exploring the supernatural involves navigating areas of uncertainty, akin to the occultist's pursuit of hidden or undiscovered knowledge.

CHAPTER 8

THE SECULARIST

Annie and Lucian visit her colleague, Dr. Elias Finch, at his secluded cabin nestled in the woods. Elias introduces them to Ponky, a humanoid robot whose movements are so fluid and responses so natural that he seems almost alive. Lucian is captivated by Ponky's attentiveness and precision, observing as the robot assists Elias with a sophisticated chemical experiment and effortlessly brews a perfect cup of coffee.

Lucian comments, "It's like he knows what he's doing." Elias shakes his head, "He doesn't know. He calculates what to do. That's the magic and the danger."

During their conversation, Elias delves into the philosophy of robots like Ponky. He explains that these machines simulate consciousness so convincingly that it's easy to overlook their lack of true understanding. "It's not magic in the supernatural sense," Elias clarifies, "but physicalist magick. The operation is real, but the consciousness is an illusion."

Annie elaborates, "Science helps us differentiate true consciousness from mimicry. Ponky is just a tool, an advanced calculator. He doesn't feel or understand; he merely processes inputs and predicts outputs. It's crucial we remember that."

Lucian frowns, "But he seems so... alive." "That's the peril," Elias responds. "People can get lost in this illusion, treating robots as companions or even deities. They forget the distinction, which can distort their reality. That's why we've decided to halt the pursuit of robotic consciousness. It's better to keep silicon devoid of self-awareness, strictly for serving human needs. Let nature retain the monopoly on true intelligence."

Despite their careful measures, rumors about Annie's team spread, igniting fear and misinformation. Politicians, leveraging fears of "soulless machines" and "godless science," file a complaint, leading to a government raid on Annie's farm. Agents confiscate the robots and detain several team members.

Lucian, witnessing the absurdity, remarks, "They're treating Ponky like he's a criminal. But he's just... an object."

Annie, shaken but resolute, reflects on the government's actions, "They're arresting tools. Calculators with limbs. It's absurdly alarming."

Summoned to a town meeting, Annie and her team face a wary crowd. They explain that their robots are meant to assist, not supplant, humans, detailing contributions to local farming, healthcare, and infrastructure that have enriched lives without undermining autonomy.

Lucian supports this view, recounting how the robots aided him in repairing his tractor and optimizing irrigation. "They're tools, not beings," he insists. "You don't arrest a hammer for performing its function."

The discussion expands to address the broader implications of government intervention and the separation of church and state. Annie's team shares experiences of financial sabotage, stalking, and military threats they've endured. They argue for a secular approach to science, which respects personal freedoms and drives innovation, contrasting this with religious overreach, which can lead to suppression and fear.

Annie states, "Separation of church and state doesn't mean prohibiting religion. It ensures no single group — whether religious or secular — can impose its will on others through force. Coexistence should be free from coercion."

Inspired by the team's bravery, Lucian organizes a protest against local politicians. He emphasizes the beneficial impact of Annie's work on the community, advocating for the protection of their rights. Gradually, more residents rally behind the "magickians," acknowledging their aid despite discomfort with the technology.

A local farmer admits, "I don't trust those machines; they seem unnatural. But Annie and her team have always been there when we needed them. That's more than I can say for the politicians."

By the end of the meeting, opinions shift. The community votes to release the robots and allow Annie's research to continue, now under public scrutiny for transparency. The ordeal fosters a deeper respect for individual rights and the ethical balance in scientific progress.

The government's suppression had forced Annie's team into a sort of hiding, which was detrimental to scientific advancement. They couldn't network or brainstorm openly as they needed to, stymieing their work. However, with the community now choosing to embrace the potential positive uses of technology and acknowledging the right to both religious and non-religious expression in Appalachia, a new chapter begins. Determining how much of Appalachia to allocate for the "magickians" remains a challenge, but it's clear they have a right to own property, just as their scientific endeavors are a form of intellectual property. Who's to say humans can't fly to the moon or cure cancer? Open science ensures that research remains transparent and deeply connected with the community, allowing everyone a chance to participate and stay informed about the innovations shaping their world.

Lucian concludes, "We don't have to agree on everything. But respecting each other's right to think, create, and live freely — that's what strengthens a community."

Secularism

The term "secular" originates from the Latin word "saecularis," meaning "of an age" or "worldly," derived from "saeculum," which refers to a "generation" or "age." In ancient Rome, a "saeculum" was typically considered to span about 100 years, marking the lifecycle of a generation.

Secularism does not inherently denote the absence of religion. Instead, it can also apply within religious contexts; for example, a Christian might not regularly attend church services or seek guidance from clergy, embodying a secular approach to faith. Secularism encompasses both non-religiosity and the broader principle that all individuals, regardless of their beliefs, should have an equal, neutral, and balanced opportunity to contribute to societal development. This framework allows both religious and non-religious individuals to engage in scientific research and environmental stewardship based on empirical evidence, objectivity, and reason.

Secularism supports religious freedom as well as the freedom from religion. A Christian, for instance, might maintain a personal relationship with Jesus Christ without adhering to church mandates. Secularism ensures that no church can compel attendance, and conversely, the state cannot interfere with one's personal faith. Thus, even in a state where atheism might be prevalent among officials, a secular Christian's right to practice their faith is protected. The separation of church and state serves to safeguard both religious and non-religious individuals, limiting governmental control over personal beliefs and lifestyles.

Secularism was pivotal during the French Revolution, where it challenged the Catholic Church's authority, leading to the concept of "laïcité," or "secularism" in French. This era, along with the Enlightenment's push for democratization and the separation of religious from political power, paved the way for secular states in the 19th and 20th centuries. Turkey, for example, adopted a secular governance model inspired by these movements.

Post-World War II, secularism advanced with the intent to reconstruct societies based on democratic values, ensuring religious freedom and promoting inclusivity that respects human rights, equity, and diversity in belief or non-belief. As András Sajó noted in 2008 in the International Journal of Constitutional Law, secularism does not advocate for a specific stance on religion's reality or its role in society. It is neither fundamentally atheist nor humanist but rather facilitates the coexistence of diverse perspectives.

The Secularist

To embody secularist beliefs, one must adopt a nuanced approach that respects individual freedom while actively participating in intellectual discourse. Secularists advocate for a society where personal beliefs are seen as a matter of personal choice, emphasizing autonomy and free will. Rather than promoting rigid dogmas or oppressive ideologies, secularists value the coexistence of a rich tapestry of ideas, fostering a harmonious environment.

For the secularist, navigating the complexities of conflicting viewpoints is an exercise in reason and rationality. While debates can be challenging, a commitment to tolerance and non-violence serves as a guiding light. Secularists uphold and defend the right of individuals to freely choose their beliefs, prioritizing personal liberty over the imposition of a singular worldview.

At the core of secularist philosophy is a firm rejection of tyranny and dictatorship. This commitment extends from personal beliefs to the broader societal framework, underscoring the necessity of engaging with those who hold differing views in a fair and open dialogue. The secularist's vision of secular freedom in society aligns with creating an environment that not only respects but also celebrates human liberty and the diversity of opinions.

Religiosity describes the extent to which an individual adheres to, practices, or identifies with the beliefs, rituals, and traditions of a particular religion. While atheism is typically defined by a lack of belief in deities, some atheists engage in religious practices for cultural, social, or ethical reasons, known as "atheistic religiosity" or "cultural atheism."

For instance, someone with Jewish heritage might participate in Jewish customs and festivals, valuing the cultural and communal aspects without holding a belief in God. Similarly, certain members of The Satanic Temple practice a form of Satanism that does not involve belief in Satan as a literal being but rather focuses on its philosophical and symbolic elements.

Atheists might also find value in religious texts or traditions for their moral teachings or sense of community and identity. Atheistic religiosity thus illustrates the diverse ways individuals relate to religious traditions, focusing on cultural, ethical, or social dimensions rather than theological ones.

Nonreligiosity refers to the absence of religious beliefs, practices, or faith. Nonreligious individuals, often called nonbelievers, include atheists, agnostics, secular humanists, and those unaffiliated with any religious tradition. This stance does not automatically imply a rejection of all religious elements; nonreligious individuals might still partake in cultural or secular aspects of religious holidays, such as celebrating Christmas for its festive or communal qualities.

Gnosticism encompasses a variety of religious and philosophical beliefs from antiquity, focusing on personal, direct spiritual knowledge

242

(gnosis). Gnostics often view the material world as distinct from a higher, divine reality, believing in salvation through understanding one's spiritual essence and transcending the physical. Gnosticism has historically been at odds with mainstream religious doctrines, emphasizing spiritual enlightenment over material existence.

In a broader sense, gnosticism can also refer to an epistemological claim of possessing ultimate, certain knowledge about truths or spiritual realities. This perspective often involves a profound realization of truths about existence, consciousness, or the nature of reality, suggesting that one can achieve a level of insight or certainty beyond ordinary human understanding.

Agnosticism is a philosophical stance that acknowledges the limitations of human knowledge regarding ultimate truths like the existence of God or the nature of the afterlife. Agnostics argue that these questions transcend current scientific or rational inquiry, remaining fundamentally unknowable. This position encourages open-mindedness and continuous inquiry, neither affirming faith in deities nor embracing atheism outright, but rather suspending judgment due to a lack of conclusive evidence.

Atheism is characterized by the absence of belief in any deity or god. Atheists do not accept the existence of divine entities, seeing no compelling evidence or rationale for such beliefs. They distinguish between observable, physical beings like humans and the concept of transcendent, non-physical deities. An atheist might engage with religious culture or ethics but lacks theistic faith.

Theism academically refers to belief in a personal, transcendent God or gods who are considered creators and rulers of the universe and are often worshiped. This term encompasses monotheism (worship of one god, e.g., Christianity, Islam), polytheism (worship of multiple gods, e.g., ancient Greek religion, Hinduism), and pantheism (divinity in all things, seen in some Taoism or Wicca practices). Theism involves a belief in a supernatural entity with attributes like omnipotence, omniscience, and benevolence, which fundamentally shapes the moral, ethical, and metaphysical outlook of its adherents.

Anostic atheism combines the acknowledgment of not knowing whether deities exist with a lack of belief due to insufficient evidence.

Contrarily, gnostic atheism holds a certain conviction that no gods exist, asserting that this belief is supported by evidence and logic.

Agnostic theism merges the belief in God or gods with an acceptance of the limits of human knowledge, suggesting that the nature or existence of the divine might be ultimately unknowable.

In gnostic theism, theism is combined with a pursuit of divine knowledge (gnosis), often involving the belief that the material world is flawed and that one must transcend it through mystical insight or revelation to achieve spiritual enlightenment.

Skepticism

Skepticism and critical thinking are pivotal cognitive strategies for navigating the complex landscape of knowledge and assessing intellectual claims. These skills transcend traditional disciplines and are particularly valuable in the realm of physicalist, secular magick, where the focus is on developing systematic, practical methods grounded in logic and empirical evidence. Skepticism involves a healthy questioning of assumptions and claims, prompting individuals to demand substance before accepting assertions as truths. Critical thinking, conversely, is the process of analyzing, interpreting, and evaluating ideas or information to achieve a more nuanced understanding of various phenomena. Together, these tools provide a robust framework for approaching the study of physicalist, secular magick with a rigorous and analytical mindset.

Skepticism is an essential intellectual stance that is rooted in inquiry and meticulous investigation. Skeptics do not take statements at face value; instead, they require supporting evidence or challenge assertions outright. This approach is especially beneficial when dealing with conflicting claims, ambiguous evidence, or situations where proof is lacking. In the context of physicalist magick, skepticism becomes a crucial tool for scrutinizing magical and magickal phenomena. For instance, a physicalist magickian might critically evaluate the purported benefits of a ritual, meticulously examining its mechanics and demanding scientific validation before accepting or dismissing its efficacy. This commitment to evidence-based scrutiny aligns with the principles of logic and critical thinking that are central to the physicalist approach to magick.

Exercise #22: Skepticism: A Simple Assessment

Is the description of the experiences subjective, open to interpretation, or does it rely on measurable, objective evidence? Consider whether the claims are based on personal feelings or perceptions, or if they can be backed by data or observable facts.

Could cognitive biases, like confirmation bias or pattern recognition, be influencing the interpretation of events? Reflect on whether personal biases might lead to seeing patterns or connections where none exist or selectively interpreting information to fit preconceived notions.

Can natural phenomena or known scientific principles explain the so-called magical or magickal occurrences? Investigate whether what appears to be supernatural or magickal might have a basis in natural causes or be explained by current scientific understanding.

Criticism

Criticism is often met with apprehension, as the prospect of having our work scrutinized can provoke feelings of inadequacy. The fear of exposing flaws or facing failure might make us resistant to constructive criticism, as we chase an unattainable ideal of perfection. However, recognizing our limitations and accepting that perfection is an illusion can transform criticism from a daunting prospect into a valuable tool for growth. Embracing criticism aligns with the fundamental idea that there are always aspects of existence, knowledge, and personal development that can be enhanced through the lenses of science and occultism. By viewing criticism as a companion rather than a threat on the path to self-improvement, we can cultivate resilience and a dedication to ongoing development.

Critical thinking is a multifaceted skill set employed to scrutinize claims and the evidence that supports them. It involves systematically breaking down arguments, identifying biases and assumptions, and assessing the validity of reasoning. Critical thinking necessitates making decisions based on facts and logic, compelling individuals to differentiate between credible and questionable sources of information. It serves as a cognitive instrument for navigating the vast ocean of data, enabling individuals to sift through content, spot logical fallacies, and make well-informed decisions. Critical thinking skills are indispensable for fostering an intellectually rigorous approach to evaluating claims and dealing with the complexities of information across various fields.

The Mechanical Spirit

The term "mechanical spirit" invokes a physicalist perspective that challenges the conventional dichotomy between spirit and body. This concept, drawing inspiration from Walt Whitman's naturalist poetry, posits that there is no inherent separation between the spiritual and the physical. In this paradigm, "spirit" is reimagined as the essence of a body, whether that body is animate or inanimate, organic or mechanical. It underscores the notion that spirit resides in the very movement, function, and evolution of matter.

Under this framework, mechanical spirits are conceptualized as entities with tangible boundaries and demonstrable functions. Even if one were to acknowledge entities like ghosts, they would be seen as possessing a form with defined limits, ultimately constrained by physical laws. The capacity to move, change, think, and exist is regarded as an attribute of the physical, highlighting the importance of integrating physics into spiritual discussions.

The traditional boundaries between the spiritual and physical blur within the concept of mechanical spirit. It suggests that the essence of spirit is intrinsically linked to the mechanics of the body, thus questioning the dualism that typically separates the ethereal from the material. According to this view, even the most intangible aspects of spirituality have a measurable, observable manifestation, encouraging a holistic and physical exploration of existence that goes beyond traditional dualism.

Secular magick and occultism serve as methodologies for embracing this perspective, offering tools for individuals to navigate and understand the interplay between the spiritual and the physical.

Exercise #23: Dissecting Spirit

Select an entity or system for dissection, ensuring it has distinct parts and functions. This could be the anatomy of a plant, the physiological systems of an animal, or a complex mechanical device like an engine or computer.

Compile or visually represent the individual components of your chosen subject. For biological entities, this might involve listing organs, tissues, and structures. For machines, identify gears, circuits, sensors, and other moving parts.

Investigate the connections and interactions among these components. Reflect on:

The relationships between parts.

How parts depend on each other.

Feedback loops that maintain or adjust the system's functionality.

Determine how each component contributes to the whole system's operation:

Trace the flow of energy, information, or materials within the system, whether it's biological or mechanical.

Consider how each part's function supports or affects the others.

How does it adapt or react to changes in its environment?

What are its responses to stimuli or challenges?

Think about the dynamic nature of its existence and how it evolves or maintains homeostasis.

From your analysis, make inferences about the mechanical spirit of your subject:

How does understanding these mechanical details enhance your grasp of the entity as a dynamic, integrated system?

What does this reveal about the concept of spirit in a physicalist context?

CHAPTER 9

THE PRACTITIONER

The Buttercream Inn buzzed with the murmur of voices, the clink of glasses, and the gentle strum of Lucian's guitar. Seated near the crackling fireplace, his hat tipped back, Lucian wore a mischievous grin, ready to spin his tale. The room hushed, the audience leaning in, eager for the magic of his storytelling.

"This one," he began, "is about the magickians and their fool of a friend — me."

Lucian recounted his initiation into Annie's team, describing the wonder of watching them fuse science, art, and technology in ways that seemed almost mystical. "I thought I'd be the comic relief," he chuckled. "But I wasn't wrong."

He shared his first blunder with Annie's solar panel project. "I crossed a wire and shorted everything," he said, laughing. "Annie just said, 'That's the cost of learning.'"

Through mistakes and perseverance, Lucian found his niche, from repairing robots in the shop to helping design solar panel installations for old Appalachian homes. "With Annie's team," he noted, "perfection isn't the goal. Showing up to learn is."

Annie's solar innovation was a game-changer. Lucian explained how these panels, with a water-based backup, ensured power even under the mountain's often cloudy skies. "Folks called them 'Annie's

Sunflowers,'" he said. "They'd seek the sun when they could, but thrive in the rain too."

These panels not only saved money but restored dignity, offering energy independence. "Annie could've made a killing selling them elsewhere," Lucian remarked. "But she chose this place, where the need was greatest. That's her essence."

Their projects, from robotics to renewables, were community-funded and valued for their efficacy. "When something works," Lucian mused, "people don't mind contributing. They take pride in it."

Lucian's highlight wasn't the tech; it was the stories. He regaled the crowd with tales of mishaps, like a robot herding goats into a churchyard or testing a solar prototype in a thunderstorm.

"I've made countless mistakes," he confessed, "but they're new each time, keeping life thrilling. Life's a story, and the best chapters are the ones with twists."

He likened himself to the Fool in tarot, stepping off cliffs not out of fearlessness but trust in the journey. "Living wild and free isn't recklessness," he said. "It's about embracing the dream—risk, error, and growth."

Wrapping up, Lucian brought his story to the present, at the Jonesborough storytelling festival. He painted a picture of the engaged crowd, the laughter, and how the mountains seemed to listen.

"Annie and the team are here, somewhere," he said, glancing around. "Probably cooking up their next big thing. Me? I'm just the fool with a guitar, glad to be part of it."

He then played a song he'd written about the mountains, the magickians, and the joy of creating something real, the audience joining in with applause and song.

As night fell, Lucian and Annie sat by the fire. "Do you think we'll ever get everything right?" he pondered.

Annie smiled warmly. "No. But getting it right isn't the point. The point is to practice, to try, to build something meaningful, even if it's not perfect."

"To the practice," Lucian toasted, strumming his guitar.

"To the practice," Annie echoed.

With the fire's glow and stars twinkling above, they shared a moment that felt timeless — one of effort, magic in trying, and the beauty of human endeavor.

The Magick of Practice

Through the exploration of storytelling, philosophy, physicalism, secularism, magic, magick, science, and occultism, a unique path unfolds—one marked by autonomy, agnosticism, mathematics, design, sensory experience, and adaptability. This tapestry invites readers to weave these diverse elements into their own narratives, creating a fusion of concepts that challenge traditional boundaries. The tools provided here act as dynamic guides, whether they are embedded within religious contexts or applied for agnostic and secular well-being. They foster a mindset that embraces physical, transformative, and creative thinking, offering a nuanced approach to navigating life's complexities. With the guidance of Jangled Jester, this approach transcends regional confines, echoing worldwide as a choice for intellectual and spiritual exploration.

Engaging with storytelling, philosophy, physicalism, secularism, magic, magick, science, and occultism on a regular basis offers a profoundly enriching life experience. Storytelling allows us to craft narratives that transform the everyday into something enchanting, while philosophy prompts us to delve into fundamental truths. Physicalism grounds us in the practical realities of the world, and secularism advocates for a fulfilling life free from the constraints of tyrannical dogma.

Magic and magick open portals to expanded consciousness, highlighting the extraordinary within the ordinary. Science provides a bastion of reason, promoting methodical exploration, whereas occultism

258

delves into the mysteries that elude conventional understanding. From these varied pursuits, three core themes emerge: curiosity, which drives us to explore the unknown; creativity, which allows us to envision new possibilities; and critical thinking, which equips us to evaluate and navigate complex ideas thoughtfully. These are the essence of Jangled Jester's vision, shaping a life of discovery and innovation.

When readers venture into this multidisciplinary journey, they transcend mere observation; they become creators, actively shaping the evolving landscape of their lives. By integrating physicalist, philosophical, scientific, and design concepts, individuals engage with the world in ways that foster curiosity, creativity, and an appreciation for the wonders of daily life. A physicalist perspective encourages a vibrant exploration of matter and forces, while philosophical inquiry invites deep questioning and reflection on life's complexities. Scientific methods offer an evidence-based approach to understanding our environment, turning the mundane into avenues for discovery. Design thinking adds a creative layer, empowering individuals to envision, reimagine, and adapt to everyday scenarios. This holistic approach nurtures an active, curiosity-driven engagement, forging a profound connection with the vastness of existence.

Exercise #24: Applying Curiosity, Creativity, and Critical Thinking in Daily Life

Start each day with a question about something in your immediate environment or something you'll encounter. Why does this plant grow here? What makes this piece of technology work?

Dedicate at least 15 minutes daily to explore an unfamiliar topic or place. This could be as simple as reading about a new scientific discovery, visiting a part of your city you've never been to, or trying a new recipe.

Pick a routine task and find a new way to do it. How can you make your commute more imaginative or your daily chores more engaging?

Engage in a weekly creative project, whether it's writing, drawing, or even designing a small experiment at home. The goal isn't perfection but exploration.

Choose a common belief or practice and argue both for and against it. This could be something as simple as the order in which you do your morning routine or a societal norm.

Before making decisions, from what to eat for dinner to how to handle a work problem, gather information, consider the evidence, and think about potential biases. Ask yourself, "What do I know, and how do I know it?"

What new questions have arisen from your curiosity?

Which creative solutions or ideas have you implemented or considered?

How has your critical thinking challenged or changed your views or decisions?

Observe how these three elements begin to shape your interactions, decisions, and personal growth. Document your journey, noting successes, failures, and insights. This not only applies Jangled Jester's vision in your daily life but also fosters a continuous cycle of learning and adaptation.

Exercise #25: Committing to Practice

Begin by understanding that mastery takes time. Define what "average" or "mediocre" means in your area of practice.

Keep a log of small achievements, even if they're not significant breakthroughs. Recognize that each step forward, no matter how small, is progress.

Each time you make a mistake, write it down. Detail what went wrong, what you learned, and how you can improve next time. This diary will serve as a reminder that errors are part of the learning curve.

Adopt the mindset that failures are lessons. After each failure, ask yourself, "What does this teach me?" rather than "Why did I fail?"

Make it a habit to ask at least one question daily related to your practice. Why does this technique work? What could be different?

Keep a journal where you note down questions that arise during your practice. This encourages ongoing exploration and learning.

Exercise #26: Transforming with Magick

Recognize magick not just as rituals or spells but as any process that instigates transformation. From personal growth to environmental changes, all are forms of magick.

Understand that not all change leads to positive outcomes. Toxic relationships, environmental degradation, or oppressive systems are examples of "black magick" in practice.

See every situation as a system of change. How do elements within this system interact to create or disrupt balance?

Like a magician, notice where perception might be manipulated or where there are misdirections in your life or society. What are the true sources of change?

Consult the laws of physics. How does matter and energy contribute to the change you observe or wish to create?

Ask why changes occur, what they mean, and what they imply about existence or morality.

Recognize the narrative arc in life events. What's the plot of change in your life or community? How can you influence it?

Venture into the unknown aspects of change. What mysteries or hidden forces are at play?

Test your theories of change. What evidence supports or contradicts your understanding of how things evolve?

Consider how changes respect or infringe upon the autonomy and rights of others. Ensure your magick promotes mutual respect and independence.

Commit to daily practice, understanding that change, like magick, requires consistent effort.

Lucian's Last Lesson

The wind howled like a living thing, tearing through the valley as Lucian clung to the rocky outcrop. The storm was unlike any he'd ever seen — electric and alive, a swirling chaos of light and shadow that seemed to fold the world in on itself. He'd ventured up the mountain to reflect on his journey, to clear his mind. Instead, he found himself face-to-face with a tempest that felt like the embodiment of every question he'd ever asked.

The storm wasn't just weather. It was a metaphor made flesh, a magickal matrix of the forces that governed the world: chaos and order, destruction and creation, fear and awe. He could feel it in his bones, the vibrations of something ancient and unyielding.

As he stared into the heart of the storm, the themes of his life flashed before him:

The fool's courage that had led him here, to the Buttercream Inn, to Annie's team, to a life where magick wasn't just an idea but a practice.

The storyteller's wisdom, weaving narratives that connected people, sparking change through shared understanding.

The practitioner's discipline, committing to the grind of trial and error, building a life of intention one small step at a time.

Lucian thought of Annie and her solar panels, the way she had turned sunlight and water into a lifeline for Appalachians. He thought of

265

the mechanic shop, where he tinkered with EVs and robots, finding joy in the messy, creative process of fixing and building. He thought of the stories he told, not just to entertain, but to inspire—to remind people that they were part of something bigger, yet still captains of their own vessels.

The storm grew fiercer, its winds pulling at him like invisible hands. A flash of lightning illuminated something strange: a book lying on the ground, untouched by the rain. It was identical to the Secular Occultist's Handbook of Physicalist Magick. His heart raced as he opened it, flipping through its waterlogged pages.

The text shimmered and shifted, revealing words he hadn't written:

"Each person is initiated by birth into the nine mysteries of this book. It is your right and responsibility to experience these mysteries, to speak and think for yourself. Yes, it will be challenging to be the captain of your own vessel, but that is what you are.

'Man is the only creature who refuses to be what he is.' Responsible. That is your first step. Be responsible.

These nine mysteries are not illusions, though they are filled with surprises and tricks. They are truths hidden in plain sight:

Stories

Philosophies

266

Physical principles

Magick

Magic

Occultism

Secularism

Science

Repetition of practice

To master them is not to conquer them, but to embrace them as the forces that shape your life. The storm will pass, but the matrix remains. Will you shape it, or will it shape you?"

Lucian felt a surge of clarity. The storm wasn't just an external force; it was a mirror of his internal struggles and triumphs. It was a reminder that while some elements of life are beyond control, there is always an opportunity to create within the chaos. He looked around, noticing the debris scattered by the storm — twigs, leaves, and shards of glass.

With trembling hands, he began to arrange the materials into a small sculpture, a representation of balance amid disorder. The act was simple, almost childlike, but it grounded him. As he worked, the storm began to subside, the winds softening, the rain easing into a gentle drizzle.

He stood back to admire his creation—a crude but beautiful symbol of transformation. He realized then that this was the essence of magick: not commanding the storm to stop, but finding a way to create meaning within it.

As he descended the mountain, the book in hand, Lucian felt a renewed sense of purpose. He would return to Annie and the team, to the stories and the work that had become his life. He would continue to practice, to learn, to fail, and to try again.

Because magick wasn't about perfection but about persistence. And in the end, that was the greatest trick of all: turning the chaos of life into something beautiful and meaningful, one small act at a time.

Putting It Into Practice

Magick is not an escape from reality; it is an engagement with it. The nine mysteries are not answers but tools, not merely solutions but formulas. Use them to shape your life, to create within the storm, and to find meaning in the ever-changing matrix of existence.

Everyone starts with stories. These narratives shape our initial understanding of the world, igniting questions, wonder, and curiosity. As we mature, we engage with nature, its resources, and the principles of physics. We are entranced by nature's illusions and mysteries, yet we also uncover layers we can measure, analyze, and theorize upon. Our consciousness is bound, imposing limits on intelligence and awareness. Yet, in the face of the unknown, a philosophy of knowledge-seeking curiosity flourishes, where even 'stupidity' becomes a catalyst for learning.

Through exploration, we bring substance into awareness, testing our ideas via experiments and the scientific method. This disciplined pursuit aligns with secularism, advocating for human freedom through voluntary assembly, minimal government intervention, and the free exchange of ideas. This confluence of principles embodies the essence of a secular occultist and a physicalist magickian.

Our framework encompasses nine chapters: the Storyteller, the Philosopher, the Physicalist, the Magician, the Magickian, the Occultist, the Scientist, the Secularist, and the Practitioner. It provides a philosophical structure for nonreligious individuals to find fulfillment, challenge, and meaning in secular art and science. Rooted in observable facts, it embraces curiosity, abstraction, and wonder without depending on deities or mystical forces. It seeks a balance between gnosticism and agnosticism, advocating for open-mindedness and critical thought.

Inject originality into the world. Initiate projects, start businesses, or launch campaigns. Join or form teams with purpose. Manifest your ideas into reality.

Venture into the vast unknown. Acknowledge that you won't grasp everything, but channel uncertainty into a journey of discovery. Let your questions lead you, transforming your so-called 'stupidity' into wonder.

Be an influential force. Friction in ideas is natural and fosters growth. Celebrate your individuality, find your unique rhythm. Define your values, share your story, and explore the essence of your being.

Inspire others to discover astonishment. Teach them to unravel the codes of the universe. Be brave in facing the unknown, guided by the integrity of scientific inquiry.

Respect the liberty of others — their minds, bodies, and property. Wake up each day prepared to apply these principles to transform your life in tangible, inventive ways.

In doing so, you embody the practice of secular occultism and physicalist magick, mastering the art of transformation through reason, creativity, and wonder. Whatever you make with these concepts, will you share your story with me, magickian?

Notes

_____

_____

_____

_____

_____

_____

_____

_____

_____

_____

_____

_____

_____

_____

_____

References

András Sajó. (2008). Preliminaries to a concept of constitutional secularism. International Journal of Constitutional Law, 6(3-4), 605–629. https://doi.org/10.1093/icon/mon018

Armstrong, D. M., Bigelow, J., & Bennett, J. (2001). Reality and Humean supervenience: essays on the philosophy of David Lewis. Rowman & Littlefield.

Asprem, E. (2016). The Golden Dawn and the O.T.O. In G. Magee (Ed.), The Cambridge Handbook of Western Mysticism and Esotericism (pp. 272-283). Cambridge: Cambridge University Press. doi:10.1017/CBO9781139027649.024

Bear, A., & Rand, D. G. (2016). Intuition, deliberation, and the evolution of cooperation. Proceedings of the National Academy of Sciences, 113(4), 936–941. doi:10.1073/pnas.1517780113

Berryman, Sylvia. (2023). Democritus. In Edward N. Zalta & Uri Nodelman (Eds.), The Stanford Encyclopedia of Philosophy. Retrieved from https://plato.stanford.edu/archives/spr2023/entries/democritus/

Benevich, F. (2017). The Essence-Existence Distinction: Four Elements of the Post-Avicennian Metaphysical Dispute (11–13th Centuries). Oriens, 45(3-4), 203-258.

Bizley, J. K., & Cohen, Y. E. (2013). The what, where and how of auditory-object perception. Nature Reviews Neuroscience, 14(10), 693-707.

Britannica, The Editors of Encyclopaedia. "Harry Kellar". Encyclopedia Britannica, 6 Mar. 2023. Retrieved from https://www.britannica.com/biography/Harry-Kellar

Butterworth, Philip. (2012). Magic in the Medieval Theater. obo in Medieval Studies. doi: 10.1093/obo/9780195396584-0004

Calhoun, C., Juergensmeyer, M., & VanAntwerpen, J. (Eds.). (2011). Rethinking secularism. OUP USA.

Carroll, J. A. (1981). The Language Game: Talismans for Language Study. The English Journal, 70(5), 83-85.

Crowley, A. (1992). Magick in Theory and Practice (6th ed.). Castle Books.

Dawkins, R. (2006). The God Delusion. Boston, MA: Houghton Mifflin Harcourt.

Delprato, D. J., & Midgley, B. D. (1992). Some fundamentals of BF Skinner's behaviorism. American psychologist, 47(11), 1507.

De Lange, F. P., Heilbron, M., & Kok, P. (2018). How do expectations shape perception?. Trends in cognitive sciences, 22(9), 764-779.

Deleuze, Gilles, and Félix Guattari. What is philosophy?. Columbia University Press, 1994.

Dijkstra, N., Bosch, S. E., & van Gerven, M. A. (2017). Vividness of visual imagery depends on the neural overlap with perception in visual areas. Journal of Neuroscience, 37(5), 1367-1373.

Drake EE. (2002). The power of story. J Perinat Educ, 11(2), ix-xi. doi: 10.1624/105812402X88650.

Einstein, A. (1905). On the Electrodynamics of Moving Bodies. Annalen der Physik, 322(10), 891–921. Retrieved from https://en.wikisource.org/wiki/Translation:On_the_Electrodynamics_of_Moving_Bodies_(1920)

Ekroll, V., Sayim, B., & Wagemans, J. (2017). The other side of magic: The psychology of perceiving hidden things. Perspectives on Psychological Science, 12(1), 91-106.

Ellis, A. (1997). Magic: The Final Fantasy. Prometheus Books.

Feather, N. T. (Ed.). (2021). Expectations and actions: Expectancy-value models in psychology. Routledge.

Francis, E. W. (2011). Magic and Divination in the Medieval Islamic Middle East. History Compass, 9(8), 622–633. doi:10.1111/j.1478-0542.2011.00781.x

Gschwandtner, C. M. (2021). Faith, Religion, and Spirituality: A Phenomenological and Hermeneutic Contribution to Parsing the Distinctions. Religions, 12(7), 476. doi:10.3390/rel12070476

Hamilton, S., & Hamilton, T. J. (2015). Pedagogical tools to explore Cartesian mind-body dualism in the classroom: Philosophical arguments and neuroscience illusions. Frontiers in Psychology, 6, 1155.

Hankins, James. (2003). Ficino, Avicenna and the Occult Powers of the Rational Soul. In La magia nell'Europa moderna.

Henry, J. (2008). The Fragmentation of Renaissance Occultism and the Decline of Magic. History of Science, 46(1), 1–48. https://doi.org/10.1177/007327530804600101

Howell, R. J. (2008). Subjective physicalism. The Case for Qualia, 125-139.

Hugh B. Urban. (2012). The Occult Roots of Scientology?: L. Ron Hubbard, Aleister Crowley, and the Origins of a Controversial New Religion. Nova Religio, 15(3), 91–116. doi: https://doi.org/10.1525/nr.2012.15.3.91

Jackson, F., Pargetter, R., & Prior, E. W. (1982). Functionalism and type-type identity theories. Philosophical studies, 42, 209-225.

Jackson, F. (1986). What Mary didn't know. The journal of philosophy, 83(5), 291-295.

Jackson, F. (1982). Epiphenomenal qualia. The Philosophical Quarterly (1950-), 32(127), 127-136.

Jonathan St B T Evans (2010) Intuition and Reasoning: A Dual-Process Perspective, Psychological Inquiry, 21(4), 313-326. DOI: 10.1080/1047840X.2010.521057

Kim, J. (1984). Concepts of supervenience. Philosophy and phenomenological research, 45(2), 153-176.

Kim, J. (2007). Causation and mental causation. Contemporary debates in philosophy of mind, 227-242.

Kocamaner, H. (2017). Strengthening the Family through Television: Islamic Broadcasting, Secularism, and the Politics of Responsibility in Turkey. Anthropological Quarterly, 90(3), 675–714. http://www.jstor.org/stable/26645759

Kuhn, G., Amlani, A. A., & Rensink, R. A. (2008). Towards a science of magic. Trends in cognitive sciences, 12(9), 349-354.

La Mettrie, J. O. (1748). L'Homme machine (Man a Machine). Paris: Durand.

Laycock, D. C., Kelley, E., & Dee, J. (2023). The complete Enochian dictionary: A dictionary of the Angelic language as revealed to Dr. John Dee and Edward Kelley. Weiser Books.

Lea Jacobs. (2011). Disappearing Tricks: Silent Film, Houdini, and the New Magic of the Twentieth Century. Journal of American History, 98(1), 225-226. https://doi.org/10.1093/jahist/jar012

Loevinger, Jane. Measuring ego development. Psychology Press, 2014.

Magic, A. E. (2016). The Experience of Magic. The Journal of Aesthetics and Art Criticism, 74(3), 253-264. doi:10.1111/jaac.12290

Marrama, O. (2019). Spinoza, Baruch. Encyclopedia of Renaissance Philosophy.

Marshel, J. H., Kim, Y. S., Machado, T. A., Quirin, S., Benson, B., Kadmon, J., ... & Deisseroth, K. (2019). Cortical layer-specific critical dynamics triggering perception. Science, 365(6453), eaaw5202.

Merkur, D. (2023, May 4). mysticism. Encyclopedia Britannica. https://www.britannica.com/topic/mysticism

Morton, P. A. (2018). Superstition, Witchcraft, and the First Commandment in the Late Middle Ages. Magic, Ritual, and Witchcraft, 13(1), 40-70. doi:10.1353/mrw.2018.0001

O'grady, P. F. (2017). Thales of Miletus: the beginnings of western science and philosophy. Taylor & Francis.

Olive, I., Tempelmann, C., Berthoz, A., & Heinze, H. J. (2015). Increased functional connectivity between superior colliculus and brain regions implicated in bodily self-consciousness during the rubber hand illusion. Human brain mapping, 36(2), 717-730.

Pitenis, A.A., Dowson, D. & Gregory Sawyer, W. Leonardo da Vinci's Friction Experiments: An Old Story Acknowledged and Repeated. Tribol Lett 56, 509–515 (2014). https://doi.org/10.1007/s11249-014-0428-7

Pitenis, AA, Dowson, D orcid.org/0000-0001-5043-5684 and Gregory Sawyer, W (2014) Leonardo da Vinci's Friction Experiments: An Old Story Acknowledged and Repeated. Tribology Letters, 56(3), pp. 509-515. ISSN 1023-8883 https://doi.org/10.1007/s11249-014-0428-7

Plato. (2000). The Republic (G. R. F. Ferrari, Trans.). Cambridge University Press. (Original work published around 380 BCE). https://ia802802.us.archive.org/20/items/PlatoTheRepublicCambridgeTomGriffith/Plato%20The%20Republic%20%28Cambridge%2C%20Tom%20Griffith%29.pdf

Plato. (2003). Republic. In G. R. F. Ferrari (Ed.), The Cambridge Companion to Plato's Republic (pp. 159-182). Cambridge University Press.

Pooley, W. G. (2023). Doubt and the dislocation of magic: France, 1790–1940. Past & Present. https://doi.org/10.1093/pastj/gtad002

Putnam, H. (1973). Meaning and reference. The journal of philosophy, 70(19), 699-711.

Putnam, H. (1981). Reason, truth and history (Vol. 3). Cambridge University Press.

Roe, C. A., & Roxburgh, E. (2013). An overview of cold reading strategies. The Spiritualist Movement: Speaking with the dead in America and around the world, 2, 177-203.

Scalabrini A, Esposito R, Mucci C. Dreaming the unrepressed unconscious and beyond: repression vs dissociation in the oneiric functioning of severe patients. Res Psychother. 2021 Aug 12;24(2):545. doi: 10.4081/ripppo.2021.545. PMID: 34568112; PMCID: PMC8451207.

Sen, R. (2018). Articles of faith: religion, secularism, and the Indian Supreme Court. Oxford University Press.

Shrum, L. J., Lowrey, T. M., Pandelaere, M., Ruvio, A. A., Gentina, E., Furchheim, P., ... Steinfield, L. (2014). Materialism: the good, the bad, and the ugly. Journal of Marketing Management, 30(17-18), 1858-1881. doi:10.1080/0267257x.2014.959

Shrum, L. J., Lowrey, T. M., Pandelaere, M., Ruvio, A. A., Gentina, E., Furchheim, P., ... Steinfield, L. (2014). Materialism: the good, the bad, and the ugly. Journal of Marketing Management, 30(17-18), 1858-1881. doi:10.1080/0267257x.2014.959

Simpson, David. (2023). Lucretius. Internet Encyclopedia of Philosophy. https://iep.utm.edu/lucretiu/

Smart, J. J. C. (1978). The content of physicalism. The Philosophical Quarterly (1950-), 28(113), 339-341.

Smart, J.J., 1981. Physicalism and emergence. Neuroscience, 6(2), pp.109-113.

Solms, M. (2019). The hard problem of consciousness and the free energy principle. *Frontiers in Psychology, 2714.

Solomon, P. R. (1980). Perception, illusion, and magic. Teaching of Psychology, 7(1), 3-8.

Stephan, A. (1999). Varieties of emergentism. Evolution and cognition, 5(1), 49-59.

Strawson, G. (2017). Physicalist Panpsychism. The Blackwell Companion to Consciousness, 374–390. doi:10.1002/9781119132363.ch2

Tarlacı, S. (2010). Why we need quantum physics for cognitive neuroscience. NeuroQuantology, 8(1).

Turing, A. M. (1950). Computing machinery and intelligence. Mind, 59(236), 433-460.

Tye, M. (1999). The subjective qualities of experience. The MIT Press.

van Inwagen, P. (2009). The mystery of existence. The Journal of Philosophy, 106(2), 692-696.

VanRullen, R. (2016). Perceptual cycles. Trends in cognitive sciences, 20(10), 723-735.

Velmans, M. (2009). How to define consciousness—and how not to define consciousness. Consciousness and cognition, 18(2), 592-602.

Welsch, W., & Singer, W. (2021). The theory of a Tetrachromatic Vision: I. Founding and Development. Foundations of Science, 26(2), 269-337.

Wolf, L. K., & Nakashima, R. (2019). Auditory Perception without Awareness: Further Evidence from a Stroop Paradigm. Frontiers in Psychology, 10, 87.